D0368580

PRAISE FOR *FEAR OF A BLACK UNIVERSE*

"The rabbit hole gets wrestled here. An old school saying applies: the more you know, the more you don't know. Dance along this read into the unknown and find out that this book may be the best ever answer to 'what is soul?'"

—Chuck D, rapper and co-founder of Public Enemy

"This book reminds me of Hawking's *A Brief History of Time*—very brief and very ambitious. It covers an enormous amount of material and offers insights not only into physics but how we do physics and who we are as physicists."

—David Spergel, winner of the 2018 Breakthrough Prize in Fundamental Physics

"Stephon Alexander has done it again. *Fear of a Black Universe* opens many dimensions—it's an endlessly stimulating, hyper-complex overview by a deeply musical scientist and mathematician. From Public Enemy's classic *Fear of a Black Planet* hip hop album and what happened before the Big Bang to how consciousness itself is woven into the fabric of space-time, this book will blow your mind. A must-read for anyone who thinks of physics and music as being inseparable."

—Paul D. Miller aka DJ Spooky

"An expansive and poetic account of not just the theory of physics, but the dreamy processes that lead to its creation, and the opposing forces that support and hinder its progress."

—Eugenia Cheng, author of $x + y$

"Einstein famously remarked that mystery is the source of all true art and science. This book explores some of the biggest mysteries of all—dark matter, dark energy, the origin of the universe, and the origin of life—in ways that are unconventional and enthralling, yet down to earth. We go on a journey with a brave adventurer for whom physics is a passionate pursuit of beauty and truth. His passion shines through on every page."

—Edward Frenkel, author of *Love and Math*

"Read this book and you'll feel awe at the grandeur and the remaining mysteries of our world, but you'll also get a hint of the human side of physics. Science is made of people and is for people; this book revives the humanist project that launched science in the first place."

—Jaron Lanier, author of *Ten Arguments for Deleting Your Social Media Accounts Right Now*

"In this courageous and provocative book, Alexander recounts his personal story of overcoming prejudice while offering a hopeful perspective for our future. Discussing the origins of his boldest ideas, from his practice as a professional jazz musician to his explorations of Jungian psychology, is especially inspiring."

—Lee Smolin, author of *Einstein's Unfinished Revolution*

FEAR OF A BLACK UNIVERSE

ALSO BY STEPHON ALEXANDER

The Jazz of Physics

FEAR OF A BLACK UNIVERSE

AN OUTSIDER'S GUIDE TO THE FUTURE OF PHYSICS

STEPHON ALEXANDER

BASIC BOOKS
NEW YORK

Basic Books
Hachette Book Group
1290 Avenue of the Americas, New York, NY 10104
www.basicbooks.com

Printed in the United States of America

First Edition: August 2021

Published by Basic Books, an imprint of Perseus Books, LLC, a subsidiary of Hachette Book Group, Inc. The Basic Books name and logo is a trademark of the Hachette Book Group.

Print book interior design by Trish Wilkinson.

Library of Congress Cataloging-in-Publication Data
Names: Alexander, Stephon, author.
Title: Fear of a black universe : an outsider's guide to the future of physics / Stephon Alexander.
Description: First edition. | New York : Basic Books, 2021. | Includes bibliographical references and index.
Identifiers: LCCN 2021001977 | ISBN 9781541699632 (hardcover) | ISBN 9781541699618 (ebook)
Subjects: LCSH: Cosmology. | Physics—Research—Methodology. | Research—Philosophy. | Research—Social aspects.
Classification: LCC QB981 .A538 2021 | DDC 523.1—dc23
LC record available at https://lccn.loc.gov/2021001977

ISBNs: 978-1-5416-9963-2 (Hardcover); 978-1-5416-9961-8 (Ebook)

LSC-C

Printing 1, 2021

In loving memory of my grandmothers,
Celisha Belfon and Ruby Farley, who taught me
how to heal with the imagination.

CONTENTS

PART I

1

ESCAPE FROM THE JUNGLE OF NO IMAGINATION

Into blinding darkness enter
Those who worship ignorance
Into as if still greater darkness
Enter those who delight in knowledge
—*The Upanishads*

We physicists have determined that over 95 percent of the matter and energy in the universe is invisible. We have branded this enigmatic stuff dark matter and dark energy; their discovery raised puzzles that shook the foundations of physical law. The gravitational effects of dark matter are observed in large halos surrounding galaxies and are critical to our current conception of how the large-scale arrangement of visible matter in the universe came to be. Likewise, so far, with dark energy, which was discovered with telescopes by measuring the accelerated expansion of the universe, it too has been the province of cosmologists, who have written about it only in reference to extraterrestrial matters and the overall shape and destiny of the universe. This is a mistake. This dark stuff turns out to play a hidden role in the visible world, including in our understanding of life itself. Dark energy resides in all empty space, not just outer space, and permeates all existence. Its quantum effects are present even in the spaces between the very atoms in our bodies. The time has come for a new Newton, to reunite the physics of the extraterrestrial with the physics of the terrestrial. Such an integration might facilitate our understanding of dark matter and dark energy, enabling a better understanding of who we are and of the cosmos in which we live.

Just as the discoveries of dark matter and dark energy shook the foundations of physics, our continued inability to unearth the identity and nature of most of the universe continues to shake them, and, consequently, it limits our understanding of our place in the universe. We still do not know much about the dark sector except that it exists; yet researchers often ascribe properties to dark matter based on presumptions that mimic known physics and are not intrepid enough. It seems to me that methodologies that might enable us to ask new questions, and find new properties or new roles for the dark in our universe, generate fear. Do we dread the dark so much that we project our fears onto the very phenomena about which we are scientifically ignorant?

Dark matter and dark energy are not the only anomalies our current physics doesn't handle. A handful of other deviations from our accepted theories of physics generate speculations that likewise trouble physics. The resolution of these anomalies may shake the foundation of what we presume to be true.

Such anomalies raise a related set of questions, one more apropos for the social sciences than the natural sciences: How does science respond to ideas that might violate our scientific norms and expectations? Does the scientific community fear embracing "dark" ideas from outsiders, especially if the ideas may not be in a form that the community is comfortable with, if they do not fit seamlessly into our theories and expected practices?

At the turn of the last century, the discovery of black-body radiation found in most objects that appeared to not emit light was a theoretical "catastrophe" for classical physics, giving nonsensical predictions that are not seen in experiments. But when German physicist Max Planck embraced the reality of the black body, he turned electromagnetic theory on its head, and the quantum revolution was born. Is it possible that the theoretical anomalies we now confront will yield a comparable scientific revolution? If so, who is likely to motivate it?

Regardless of our ability to create the most abstract mathematics and come to know truths beyond our five senses, as humans we are limited by our social and psychological conditioning. In this book, we will go beyond the current conceptual and scientific-sociological paradigm into uncharted and sometimes risky conceptual territories. What lurks beyond the black hole singularity in our galaxy and before time existed at the big bang? How did cosmic structure emerge from a chaotic and featureless evolving early universe? What is the role of dark energy and dark matter in the universe? Is there a hidden link between the emergence of life and the laws of physics? These are questions on the boundary of what we know; answering them may call into question the theories that constitute our knowledge. If we are to answer them, we must ask whether the scientific community is able to incorporate into its activities nontraditional members, outsiders more likely to see beyond our current theoretical horizon; further, is the scientific community, as it is now structured, able to empower these outsiders to break new ground?

I want to bring you with me as I try to take on some of these questions. To do this effectively, I will provide both the necessary background and the conceptual tools needed to understand a bit of established physics. My discussion is based on three fundamental principles that underlie all known physics; a grasp of them will enable us to understand some of the problems at the borders of what physicists think they know and understand. I will be frank, sometimes controversial, and deliberately engage in some of my own wild speculations. This is not just a book about what we know in physics, but a book that explicates the frontiers of physics, a book about how physics is done.

Much that is taught and written about physics expresses what we know already. This book presumes that the process of *doing* physics is different than the process of *learning* the physics we know already. The first explores what we do not yet know, while the second transmits what others have learned previously. Here, while some of the latter is necessary—there are certain things that must be shared—our focus is on how we might think about what we do not yet know.

Crucial is a simple insight: a responsibility of physicists is to apply what we know already in new areas of inquiry, to transform and extend our knowledge. Great teaching in physics helps us to do physics, not simply learn the physics we know already. This means learning physics has to enable us to work at the boundary of what we know or, in rare cases, to go beyond those boundaries, or even reconstruct the very framework of our knowledge. If this book is successful, it will help you understand what it means to be creative in theoretical physics.

Often when I am stuck on a problem—of the physics variety or the personal one—I make a pilgrimage to the northern coast of my birthplace, Trinidad and Tobago. There is something that feels unspeakably out of body about trekking through the lush sixty-mile stretch of the deep green mountain range overlooking Las Cuevas

Bay. I hike up the winding paths to the top of a hill overlooking the ocean, the tropical jungle sounds looming behind me and the rhythmic crashing of the crystalline emerald crescent waves sounding below. Surrounded by nature, beautiful and primordial, I am often surprised to find new insights into my problems.

One day not so long ago, I found myself getting nowhere on a research problem. I headed back to the jungle to look at the sea. While I was there it dawned on me—not the solution to the research problem, but the realization that during two decades of scientific research, I had been unconsciously dodging my original reason for becoming a physicist: to make a meaningful scientific discovery. I realized I feared failure and the professional risk failure entailed. The ability to maintain a scientific career is driven by, among other factors, your reputation among your peers and familiarity with your work. Penalties await those who are perceived as a "crackpot" or who speculate too much. I knew that some of the ideas that interested me, such as the connection between consciousness and quantum mechanics, would make me vulnerable to stigma and potentially stump my career.

In theoretical physics research, there is a sense of dissatisfaction, a belief that we have not been able to break new ground in the same way that led to the quantum and relativity revolutions early last century. It's not to say that people aren't trying to address their dissatisfactions; a handful of papers are posted every day on an online global archive of physics research called the Archives, and oftentimes these papers offer new approaches to unsolved mysteries. Despite this, there's not much feeling of progress. Why is this? Is it because these problems are too hard for us? Or is it that in the search for the truth, some scientists are afraid to look at uncharted or forbidden territories, afraid because there may be penalties, reputational and professional, for stepping outside accepted paradigms? I think that it's the latter. In this book, I will provide my thoughts and reflections; I will take some risks, hoping that we learn something significant along the way, whether I am right or wrong.

As a Black physicist, this potential strength—that I am brimming with ideas, my capacity to generate speculative thinking—can be an impediment. Black persons in scientific circles are often met with skepticism about their intellectual capabilities, their ability to "think like a physicist." Consequently, my exploratory, personal style of theorizing, when coupled with my race, often creates situations where my white colleagues become suspicious and devalue my speculations. I have navigated a career in physics in spite of these racial and sociological prejudices, and, given both my personality and my predilections, I continue to march ahead, sharing my conjectures, which, at least sometimes, are theoretically fruitful. This book will be no exception.

During my time of self-reckoning in Trinidad, I decided to devote the majority of my research efforts to working on some of the big mysteries in physics. To do so effectively, I would have to bring my entire being to how I do physics, which meant engaging in improvisational and wild speculations. When you meet me in person, it is clear that I am volcanic with ideas, most of which turn out to be wrong, while some, even among those that are "wrong," are fruitful and worth pursuing. Underlying these ideas is a latent foundation, the theoretical and technical tools of my trade.

Physics is a social activity, and like all social activities it is regulated by norms. Practitioners are expected to conform to these normative expectations, and they are sanctioned negatively when they violate them. Too often the expectations of what it means to do "good science" become confused with specific theoretical orientations, which means that practitioners in subdisciplines are expected to uphold specific theoretical arguments. This is desirable insofar as it rules out ideas like flat-earthism and others that make no sense scientifically. Sometimes, however, this expectation of conformity stifles innovation and progress. Some scientists are reluctant to explore ideas outside the expected paradigm because they will be punished if they do so, which means that paradigm-shattering theories can be inhibited from emerging.

We need to distinguish clearly between the values and norms that regulate scientific activity and those that demand conformity to a particular body of theory, a particular paradigm, within a scientific community. Both are constituted socially, but the latter obligations can restrict our creativity, our ability to constitute new theoretical orientations. It is crucial, however, to recognize that our theoretical arguments must be regulated within and evaluated through the application of scientific values, the values of cognitive rationality. Very simply, this means that our theoretical arguments must be logically coherent and empirically warrantable. Not every "creative idea" may be turned into viable physical theory. In fact, the likelihood that any one of us will create a new paradigm because we have violated the norms regulating activity within the standard paradigm is very slim. No one can do so, however, without violating these norms.

I want this book to serve as a source of inspiration and encouragement for individuals who feel disenfranchised and unwelcome in our scientific communities, people who are sometimes, or often, made to feel that they are not valued as contributors to the scientific endeavor. So as much as this is a book about my reflections on the state of physics, as theory, I also reflect on and analyze both the sociology of science and my own experiences to argue for the efficacy of outsiders' presence and perspectives in scientific communities and inquiry. The path to becoming a scientist poses challenges for everyone. In shedding new light on the social dynamics of science, and simply sharing our stories, we can see how some of the challenges outsiders face can inspire them to make significant scientific contributions. I hope to convince my readers that diversity in science is not simply a social justice concern, but that it enhances the quality of the science we accomplish.

Many of the theoretical physicists of my generation were inspired by the golden era in the first half of the twentieth century, when the likes of Albert Einstein, Richard Feynman, Paul Dirac, Emmy Noether, and Wolfgang Pauli, to name only a few of our idols, gave birth to quantum field theory and general relativity. These theories have been spectacularly confirmed, and they are responsible for most of our modern technology.

One of the essential tools that Einstein and Erwin Schrödinger employed in discovering the equations and fundamental laws of relativity and quantum physics was "thought experiments": mental visualizations, or imaginations of physical happenings that are impossible to carry out in terrestrial settings or with current experimental techniques. Some of the famous ones include Schrödinger's cat and Einstein's vision of riding on a beam of light. These visualizations, when articulated as mathematical equations, led to solutions that predict the behavior of the semiconductor devices that drive powerful computers, including the smartphones that are part of our everyday lives.

When I first learned how the greats managed to make these discoveries, it seemed as if some mental wizardry were at work, a wizardry that has been overlooked by my generation and our teachers. Theoretical physics has grown to become extremely mathematical, and while mathematics is a necessary and powerful tool, I realized that if I were to have a shot at making an important discovery, I would have to find my way to acquire a bit of that wizardry, the intuitions leading to the theoretical insights that lead to mathematical equations (intuitions not derivative from those equations).

As a young student taking introductory courses in physics, I had the impression that physics was a jungle of countless equations and intricate theories. The task, or so it seemed, was to digest and apply them. Even decades later, as a researcher in theoretical physics, it dawned on me that my colleagues and I were lost in that same jungle. The mentality required to work through problem sets made the handful of problems in cosmology and particle physics that seemed important also seem insurmountable. We did not even know the right questions to ask.

At Las Cuevas Bay, after gazing at the waves for some time, I had an epiphany: Who better to help us address our questions than Einstein himself? What if we had a bird's-eye view of the jungle of physics from which we could see the origins of the theories and the interconnections between the laws that give rise to (and constrain) them? Would this perspective facilitate our attempts at reworking these theories to better address our contemporary questions? Could we turn from calculating, boring physicists to brave adventurers, imagining worlds no one else had seen before?

During my time as a postdoctoral researcher in theoretical physics at the Stanford Linear Accelerator Center (SLAC), I received a surprising letter from the National Geographic Society. I wondered if I owed them a payment. Instead, the letter congratulated me on being selected as a National Geographic Emerging Explorer. I was both elated and confused. I had never applied to be an explorer, nor did I think of myself as one. When it turned out that it was not, in fact, a mistake, I was deeply honored and did not say no to the monetary prize and subsequent trips to National Geographic headquarters to meet other explorers I admired. For example, I had always wanted to meet ethnobotanist Wade Davis, whose book was the basis for one of my favorite horror movies, *The Serpent and the Rainbow*.

All explorers were invited to a fundraiser and to celebrate the seventieth birthday of the Society's president at the time, Gil Grosvenor.

There were many impressive people there, and I quickly started to feel a little out of place. Among the newly elected explorers was an underwater cave diver who could contort his body to fit into intricate caves for hours, hundreds of feet under the ocean. There was a woman who lived among lions in the Serengeti, and a man who explored and lived in Antarctica for extended periods. At the fundraiser, each explorer was placed at a dinner table with a group of potential donors, to entertain them. After we introduced ourselves, one disappointed donor said to me, "You don't hang out in the jungle? You don't fly airplanes? Why did they make a theoretical physicist an explorer?!"

I didn't want the donor to feel duped. So as a good spokesperson for National Geographic, I responded with conviction: "I explore the cosmos with my mind." I went on to explain how the worlds that cosmologists explore are even more extreme than explorers on Earth, so extreme that we are forced to explore them in our imaginations. I went on to explain that Einstein explored the nature of space-time, and this led to the ultimate prediction and discovery of a supermassive black hole at the center of our galaxy. Try exploring that physically! Some donors were interested, but others wanted to hang out with a "real" explorer.

Despite the drama, that night got me thinking about the similarities between physical and mental exploration, about the extreme places theoretical physicists must explore to make progress. These mental explorations are the fuel for discovering and clarifying physical theories; they are the domain of Einstein's notion of principle theories.

In 1914, soon after his revolutionary discoveries in quantum mechanics and relativity, Einstein gave an address to the Prussian Academy of Sciences in which he discussed his strategy for discovery in theoretical physics. "The theorist's method involves his using as his foundation general principles from which he can deduce conclusions," Einstein said. "His work thus falls into two parts. He must first discover his principles and then draw the conclusions which follow from them."

Einstein could perceive a hidden reality, where time and space could slow down, speed up, bend, and even cease to exist, a reality that transcends the limits of our daily perceptions, a reality that makes no sense to us when we are thinking commonsensically. Surely there are still new levels of reality that are hidden, and like Einstein we ought to be curious to know what lies beyond our current (commonsensical) understanding in physics.

As a student, I had mistakenly thought that physics was driven mostly by mathematics and logical reasoning. Einstein's conviction was that principles are the driving force behind new discoveries, while mathematics is necessary to make physics precise, to inform the clarification of the principles, to explain and clarify our characterizations of how we conceptualize phenomena, and to make predictions. In short, math is not enough; it is a tool. The important question is how does one come up with new principles? Einstein answers: "Here there is no method capable of being learned and systematically applied so that it leads to a new [principle]. The scientist has to worm these principles out of nature by perceiving in comprehensive complexes of empirical facts certain general features which permit of precise formulation."

He was saying that a scientist should make connections and see patterns across a range of experimental outcomes, which may not be related to each other in an obvious way. Once the scientist ekes out these patterns, she makes a judgment call as to whether a new principle of nature is necessary. But this is misleading. Facts are statements about phenomena, but they don't exist on their own; they are always conceptualized, which means that they are, if only implicitly, constructed theoretically. Experiments allow us to answer theoretically constructed questions. Theory tells us what "facts" to look for.

As an adolescent Einstein was free to play in his father's electrical company in Pavia, Italy. This play fertilized his imagination; it enabled him to envisage what he would experience if he could catch up to a light wave. His process of "worming" out these principles entails visualizing phenomena that are not directly accessible to our senses or current experiments. It eventually enabled him to formulate theories that told us what we would find and helped us to understand where we might look to find it.

How did Einstein know when to postulate his theory of relativity? How, aside from his natural-born genius, was he able to arrive at his principles? I found part of the answer in a lecture he gave at Oxford University in 1933. "[The discovery of principles] are free inventions of the human intellect, which cannot be justified either by the nature of that intellect or in any other fashion *a priori*," he said. But what does Einstein mean by this? Sometimes to get around a scientific problem, one must consider possibilities that defy the rules of the game. If you don't enable your mind to freely create sometimes strange and uncomfortable new ideas, no matter how absurd they seem, no matter how others view your arguments or punish you for making them, you may miss the solution to the problem. Of course, to do this successfully, it is important to have the necessary technical tools to turn the strange idea into a determinate theory.

When I told the donors at National Geographic that I explored the cosmos with my mind, I wasn't jiving. Those theoretical physicists who explored with their intellect, making "free inventions," sounded to me like masters of improvisation. Einstein gave me the hall pass to continue my free inventions. But, like Einstein did, we must first look to the fundamental principles underlying modern physics and use them to explore some of the big mysteries physicists face. In the pages that follow, we will engage in free inventions, trying to cook up some new physics while journeying through some of the biggest mysteries at the frontiers of cosmology and fundamental physics. While some of the ideas presented in this book are debatable and speculative, I hope that it nonetheless provides not only insight into how a theoretical physicist dreams up new ideas and sharpens them into a consistent framework but also, perhaps, the inspiration to think of your own big ideas.

2

THE CHANGELESS CHANGE

> A new idea comes suddenly and in a rather intuitive way. That means it is not reached by conscious logical conclusions. But, thinking it through afterward, you can always discover the reasons which have led you unconsciously to your guess and you will find a logical way to justify it.
>
> —Albert Einstein

After many years spent developing my skills and ideas until they were good enough for publication in physics journals, I finally published my first independent paper in the *Journal of High Energy Physics*. My article made an iconoclastic claim: Einstein's cherished idea of a constant speed of light could be violated in the early universe if our actual universe were a three-dimensional membrane orbiting a five-dimensional black hole. If this sounds like gobbledygook to you, in hindsight, it is. But twenty years ago, such subject matter was typical of what theorists worked on as they were trying to integrate cosmology and string theory. I was especially proud that the months of calculations I performed within the framework of string theory provided these new solutions.

And so there I was, excited to give my first professional talk at a picturesque university nestled in the mountains of Vancouver, Canada. They were my calculations I was going to talk about, so I knew them inside out, which contributed to my air of overconfidence. It didn't last long. Within five minutes of my talk's beginning, I was blasted with questions that soon transformed into a flood of criticism. Attempts to continue my talk ricocheted against random comments, delivered with a tone of unfriendliness, from the audience, attacking the premise of the talk: "Why should we believe our universe is a brane rotating around a 5D black hole?" I couldn't help but feel unwelcome and alienated. By the middle of the talk I stood dejected, my fears of not being accepted as a peer erupting to the surface of my mind. Just because your paper gets published doesn't mean that you will get into the club of physics. That day it felt obvious I hadn't.

Then came a voice from the back of the room. The speaker was a distinguished Indian physicist in his seventies decked out in a well-groomed tweed suit. As soon as he began to speak, everyone shut up, as if a demigod commanded his minions to silence.

The old man stood up and said, "Let him finish! No one ever died from theorizing."

It was the biggest lesson with the fewest words the audience and I could have learned about the art of theoretical physics. Those words would stay with me throughout my life as a theoretician. I took the old man's admonition as a reminder to never be afraid of even the most absurd ideas, and to even embrace them. I finished my talk without further interruption and even got a round of applause afterward. Did I take my theory seriously years later? No, but the exercise of journeying into a theoretical territory and then journeying back has proven time and time again to be useful in surveying what's possible and, hopefully, what describes and predicts the real universe. That moment was pivotal in my life and how I would engage the art of theoretical physics for the next twenty years.

A year after that talk, after many failed attempts, I landed a job as a postdoctoral researcher in theoretical physics at Imperial College in London. The department had been founded by Abdus Salam, who, along with Sheldon Glashow and Steven Weinberg, would win the Nobel Prize in Physics for discovering a unified theory of the weak nuclear interaction with electromagnetism. I was excited to be following in Salam's footsteps along the road to becoming a research physicist. Yet somehow, despite my excitement, I quickly realized that road was not what I thought it would be.

At Imperial, weeks and then months of work could pass with little to show for it. If an idea did come to me, I invariably and quickly discovered that someone had already developed and published it. If I was performing a calculation, I would often hit a roadblock and have to learn new mathematical techniques in order to tackle it. By the time I learned the new math, someone would have already hit the finish line and published the result before me. These experiences forced me to wake up from my theory dreams to a reality in which the prospect of becoming a scientist seemed dim. My contract was for two years, but I relegated my expectations to another career, perhaps going on the road as a jazz musician or teaching high school physics, both admirable things to do. I would continue trying my best, but ideas simply weren't coming, and I would continue to fake it and keep these frustrations to myself. I had everyone fooled.

Then one seemingly uneventful day, horror hit me. I received an email from our theory group administrator that simply said: "Professor Isham would like to speak with you." I turned white like a ghost. Chris Isham was the head of our theory group and I feared that he had figured out that I was a fake. Everyone in our group revered Isham for his exceptional abilities in quantum gravity and mathematical physics. He was a tall Englishman with dark hair and piercing eyes and who walked with a slight limp. Like his friend and classmate Stephen Hawking, Isham suffered a rare neurological disorder that kept him in constant pain. I had kept away from him in fear of letting some gibberish slip out to ignite his physics bullshit detector. Now I suspected and feared that he had figured it out on his own, and my day had come to face him.

I decided to do a little preparation and read one of Isham's papers. Perhaps I could appease him. To my dismay, many of his publications involved some of the most advanced concepts in math and physics, with inscrutable names such as topos theory, quantum logic, C-star algebras, and so on. I finally found a paper that he'd written two decades ago that I could grasp. It was about the behavior of quantum particles with half-integer spin, called fermions, in an expanding gravitational cosmology. Electrons, quarks, neutrinos, and most matter are examples of fermions, so it might seem a safe topic. Still, I set off nervously for the meeting.

I tensely walked into his large office filled with books, incomprehensible equations, and diagrams. On his desk an oddly placed small statue of an angel faced a visitor. After a brief hello, Isham didn't waste time.

"Why are you here?"

I kept it real. "I want to be a good physicist."

To my surprise, he said with a serious demeanor, "Then stop reading those physics books!" Then he pointed to an isolated bookshelf. "You see those books over there? They are the complete works of Carl Jung. Do you know that Wolfgang Pauli and Jung corresponded for decades? And Pauli's dreams and analysis were key to his discovery of the quantum exclusion principle."

Isham revealed that he had been studying Jung over the last fifteen years and had trained himself to do calculations in his dreams. I couldn't believe that I was hearing this from one of the master mathematical physicists on the planet. Then he had a eureka smile and said, "You know what? How about you come to my office once a week? Write down your dreams and tell me about them." He suggested that I read Jung's Volume I, book 9 entitled *Aion: Researches into the Phenomenology of Self* as well as *Atom and Archetype*, a collection of two decades of letters between Jung and Pauli. At first, I was skeptical of the experiment. But I was also feeling isolated in the theory group, and Isham's invitation to talk about my dreams was an opportunity to spend quality time with one of my physics idols.

Our weekly discussions started with me telling him about random dreams that had no apparent relation to physics, such as those about past relationships that continued to taunt me. During our time together, Isham would share his perspectives on some of the mysteries that our field faced. One of those was the problem of time in quantum gravity. While our physical (and psychological) experience of the flow of time is taken as fact, time disappears in the equations of quantum gravity. Isham worked on this problem and was a proponent of a new notion of time called internal time. It was no surprise to me to learn that these ideas were inspired by his exploration of psychology and mysticism.

As the weeks passed, I told Isham about what I thought was a trivial dream. In Jungian philosophy, dreams sometimes allow us to confront our shadows with the appearances of symbols called archetypes. I saw one here. I was suspended in outer space and an old, bearded man in a white robe—it wasn't God—was silently and rapidly scribbling incomprehensible equations on a whiteboard. I admitted to the old man that I was too dumb to know what he was trying to show me. Then the board disappeared, and the old man made a spiraling motion with his right hand. Isham was captivated by this dream and asked, "What direction was he rotating his hands?" I was baffled as to why he was interested in this detail. But two years later, while I was a new postdoc at Stanford, I was working on one of the big mysteries in cosmology—the origin of matter in the universe—when the dream reappeared and provided the key insight to constructing a new mechanism based on the phenomenon of cosmic inflation, the rapid expansion of space in the early universe. The direction of rotation of the old man's hand gave me the idea that the expansion of space during inflation would be related to a symmetry that resembled a corkscrew motion that elementary particles have called helicity. The resulting publication was key to earning me tenure and a national award from the American Physics Society. Chris Isham's method proved to work for me. But he and I weren't alone here. It turns out that some of the biggest breakthroughs in science were inspired by dreams, including Einstein's theory of relativity.

Beginning when Einstein was a teenager hanging out in his father's electric lighting company, he would play with imaginations about the nature of light. He would try to become one with a beam of light and wondered what he would see if he could catch up to a light wave. This matter found itself in the playground of Einstein's subconscious and revealed a paradox in a dream. It is said that Einstein dreamt of himself overlooking a peaceful green meadow with cows grazing next to a straight fence. At the end of the fence was a sadistic farmer who occasionally pulled a switch that sent an electrical current down the fence. From Einstein's birds-eye view he saw all the electrocuted cows simultaneously jump up. When Einstein confronted the devious farmer, there was a disagreement as to what happened. The farmer persisted that he saw the cows cascade in a wavelike motion. Einstein disagreed. Both went back and forth with no resolution. Einstein woke up from this dream with a paradox.

In the account of Einstein's dream, and other accounts of the role of dreams in creative work, such as music, science, and visual art, there is a common theme: a paradox is revealed through imaginations that are contradictory in the awake state. It's as if the mind's eye can access an intuition beyond the waking state and not restrict our imaginations to the self-editing that our conditioning might impose during the waking state (unless you're a great daydreamer). Perhaps dreams are an arena that can enable supracognitive powers to perform calculations and perceptions of reality that may be incomprehensible in our wake state. In my case, my paradox was making an equivalence between incomprehensible equations presented by the bearded man and his counterclockwise whirling hands. This counterclockwise motion turned out to summarize the mathematics that was obscuring the underlying physics to be unveiled.

My preoccupation with equations as the way to access deeper physical reality was confronted by this paradox. I discovered years later that I was not alone. My friend, virtual-reality pioneer, composer, multi-instrumentalist, and author Jaron Lanier told me that in his pre–virtual reality days, he used to hang out with Richard Feynman at Caltech. During that time Feynman was experimenting with other ways of doing physics, including using his body to intuit physics. This got the young Lanier thinking about how the human body could interact with computers in new ways and inspired what would become VR. Einstein's paradox asks how it could be possible for both realities of light to be true. He ultimately resolved that paradox, and in doing so would uncover one of his most fundamental principles, one that was key in discovering not only his theories of special and general relativity but also the nature of all known four forces!

These thought experiments with light also led Einstein to other paradoxes, such as in James Clerk Maxwell's theory of electromagnetism, which mathematically describes the motion of light waves. In this case, Einstein found inconsistencies that forbade absolute rest for a moving wave of light (an electromagnetic wave). In his groundbreaking 1905 paper *On the Electrodynamics of Moving Bodies*, Einstein resolves this inconsistency by elevating the role of principles in physical law. During Einstein's time, physicists assumed that the universe is filled with a substance called the luminiferous ether that light could move through, like water waves moving through the medium of water. Mainly because of his intuition that an electromagnetic wave of light could never come to a complete stop, Einstein gave up the ether. Based on years wrestling with a handful of conundrums with electromagnetic theory, Einstein made an intuitive leap and postulated the principle of the invariance of light:

The Principle of Invariance of Light: *The speed of light in empty space is the same for all inertial observers regardless of how fast they are moving relative to each other.*

It's worth saying a little about some of the physical reasoning that justified Einstein's adoption of this principle. In his groundbreaking paper *The Electrodynamics of Moving Bodies*, Einstein finds a paradox in Maxwell's equations that describe electric and magnetic fields. Both electric and magnetic fields exert forces on charged particles. It was well known since Galileo's time that mechanical forces would be unchanged (invariant) for observers that are moving at constant speeds relative to each other. However, Einstein found that this was not the case for Maxwell's equations. For example, different frames of reference would give different physical results for electric and magnetic forces. Einstein realized that this could be fixed if the electric and magnetic fields would also change into one another depending on the frame of reference.

Visible light is a manifestation of electromagnetic waves at a given set of frequencies that our eyes can detect. Einstein realized that if electric and magnetic waves traveled at a constant speed, the electric and magnetic force laws would remain intact. Einstein found that the transformations that related these moving observers, which are known as Lorentz transformations after the mathematician who first found them, did something surprising: they warped spatial and temporal measurements of relative observers. The consequence of Einstein's insight is that we know electromagnetic waves are unified in four dimensions, three spatial and one temporal. And this relativity extended to all known laws.

Einstein realized that if time depended on the observer's velocity it would be possible to retain the invariance of light in different moving frames. A similar argument was made for the length of objects in relative frames of reference. Therefore, both time and the sizes of objects are no longer absolute but are subject to change depending on how fast one is moving relative to another inertial frame of reference. The lesson here is that the invariance of the speed of light required time and space to be relative—we need relativity to have absolutes.

In our everyday experience, we simply do not experience time slowing down for us when we move faster. This is because the velocities we are used to are not large enough to see the relativity of time and space. Nevertheless, Einstein's new theory made these bold predictions, and it was up to experimentalists to find a way to put them to the test. Not far from where I lived in Hanover, New Hampshire, when I was a member of Dartmouth's physics department, the phenomenon of time dilation was finally confirmed by David Frisch and James Smith, who attacked the problem with the help of a particle called the muon. The muon is the cousin of the electron. Like the electron, a muon has a negative charge that we describe as −1, but there is a critical difference: a muon has two hundred times the electron's mass. The muon is also unstable and will decay into an electron at a fixed rate of two-millionth of a second. Therefore, we can treat the decay of a muon as a standard clock. In 1941 Bruno Rossi and David Hall made use of the serendipitous fact that muons travel to Earth from outer space close to the speed of light to get a rough estimate of how moving at such a high speed dilates time. They measured the half-life of these rapid muons at high altitude and found, as Einstein had predicted, that these muons lived longer than muons that were at rest. In 1963 Frisch and Smith made much more rigorous measurements of the speed of extraterrestrial muons, which enabled an even more precise confirmation of how high speed warps time. These observations were refined and continued to be confirmed with the precise numerical predictions of Einstein's theory of special relativity. But there is something even more special about special relativity.

Invariance isn't just an empirical fact. The core idea behind invariance is based on the symmetry properties of geometric objects. The key insight in Einstein's invariance principle is that physical entities like space-time and the other forces are in direct correspondence with geometry. It still baffles mathematical physicists why geometrical structures are linked with space-time and the building blocks of matter; no one has expressed this better than Nobel laureate Eugene Wigner, who wrote that "The miracle of the appropriateness of the language of mathematics for the formulation of the laws of physics is a wonderful gift which we neither understand nor deserve."

Consider the transformations of a square that leave it invariant. These are its symmetries. Rotate the square by ninety degrees and it looks the same. Rotate it again by ninety degrees and it still looks the same. Ultimately there are four successions of ninety-degree rotations that would leave the square looking the same. We can also flip the square about its horizontal center, and it looks the same. We can flip it about its vertical center and it also looks the same. We can flip the square across a diagonal line in two different ways and it looks the same. Therefore, there are eight different ways we can change the square and it looks the same. Said another way, there are eight transformations that leave the square invariant. The square is geometrically described as a two-dimensional object that has an eightfold symmetry. If I wanted to communicate with aliens that could not see, but knew numbers, I could describe many geometric objects by their symmetry transformations.

So, invariance captures an idea that science writer K. C. Cole describes as the differences that don't make a difference. In the case

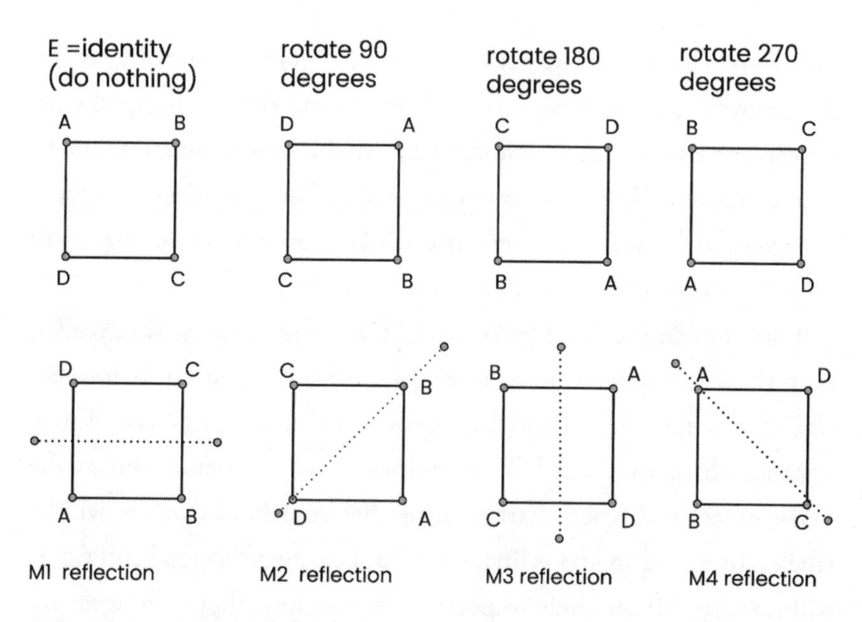

FIGURE 1: The eightfold symmetry transformations that leave the square invariant.

of special relativity, I can move as fast or slow as I want, and the speed of light will not change. And this means that we can identify an underlying geometric symmetry. What would a world like that look like geometrically?

In special relativity Einstein found that there is a symmetry in all inertial coordinate systems that leaves the speed of light the same. The key to being able to enforce that the speed of light is the same in different inertial coordinate systems is to find the correct object, or observable, to relate coordinate systems to each other. The observables that we often use are distances and time. In three-dimensional space, distances alone are not sufficient because the position of a moving object can also have a direction. The mathematical object that describes distances with a direction is called a vector. Einstein realized that he could unite space and time into a four-dimensional vector, a space-time vector. This vector has a length and direction in both space and time. But to do this, we must put the time dimension on the same footing as a spatial dimension. The hack: multiply time by a velocity and we get a dimension of length.[1] In this way the time dimension resembles a spatial dimension. Being equipped with a four-dimensional space-time vector enables us to relate them to each other in different coordinate systems. These relations or space-time vectors between different coordinates are exactly analogous to rotation transformations of points on a circle.

Consider an idealized Ferris wheel ride, where one person, Kolka, is on the Ferris wheel and another, Jim, is on the ground. Before the ride starts, both Kolka and Jim seem to share the same coordinate system, which they can label as points on an x-y plane. But as the Ferris wheel and Kolka start rotating, her coordinate system will be labeled by a rotating coordinate system. But even though both riders will disagree about their respective coordinates, they will agree on the invariant, which is the length of the radius of the Ferris wheel. In the same spirit, Einstein's theory of invariance relates coordinates that differ from each other by relative velocities but retain their four-dimensional "radius" that remains unchanged by all observers. The compromise for preserving the four-dimensional radius is that observers that move at velocities comparable to the speed of light will experience the rate of time slowing down and lengths shrinking relative to observers that are at rest.

The four-dimensional space equipped for four-dimensional vectors was invented by Hermann Minkowski, who was inspired by his former student Einstein and realized that the system of relativity had an underlying geometry that is a four-dimensional space-time continuum that had a symmetry that left the speed of light as an invariant. The constant speed of light is formulated as the geometry of a cone embedded in a four-dimensional space. In this geometry a ray of light traces out the cone, and the slope of the cone represents the constant velocity. And just like the radius of a sphere remains the same length regardless of how one moves on the surface of the sphere, in this four-dimensional space-time the light cone would remain the same no matter what frame of reference is used.

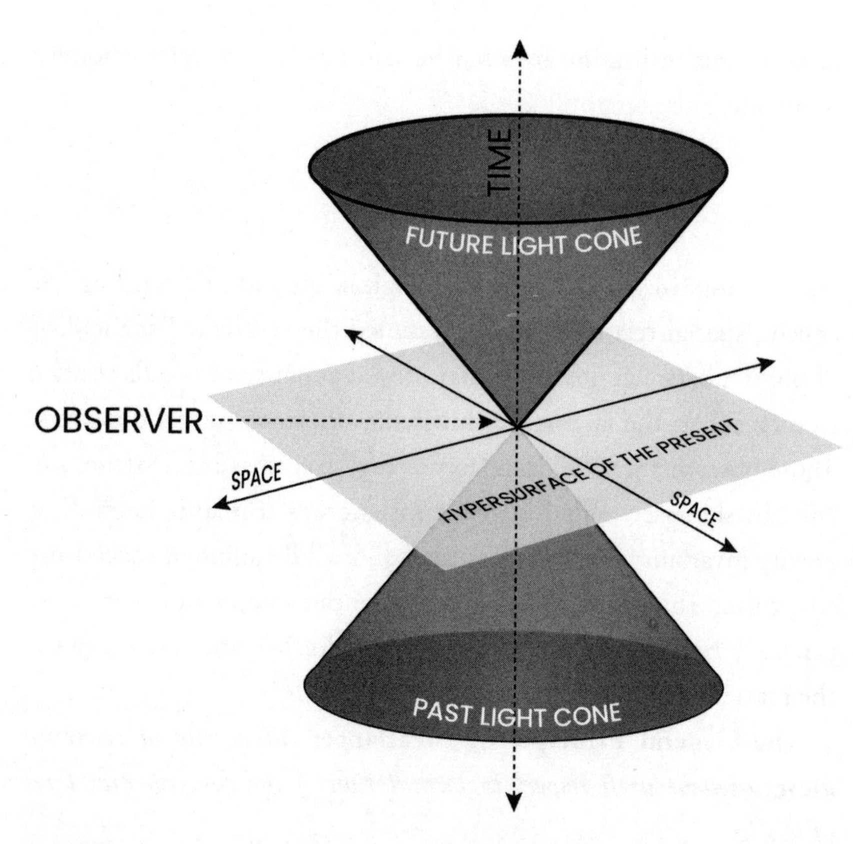

FIGURE 2: The light-cone structure of Minkowski space-time.

The triumph of special relativity was based on a simple and deep new principle of invariance, but the principle was so deep and universal that importance of the idea of invariance did not stop with special relativity. As we will soon see, the invariance principle became the master organizing principle behind all the physical forces, from all nuclear forces to gravity. Einstein further hypothesized that the laws of physics should not care about any relative state of motion, including those in which observers are accelerating relative to each other. If you are spinning in a rocket ship out in space or sitting still on a planet, the laws of physics will function identically. This led Einstein to perceive a reality beyond our five senses to discover a new law of gravity, the general theory of relativity, by deepening his invariance principle to observers undergoing any type of motion. He came to this realization by what he said was "the happiest thought of my life."

General relativity is in hindsight a logical, but radical, step from its cousin, special relativity, which assumed the invariance of the laws of physics between observers that moved at constant speeds relative to each other: the jargon for this is that they are inertial observers. Those observers were all in reference to a fixed space-time structure, Minkowski space. But Einstein's insight was to make the laws of gravity invariant for observers situated in wildly different space-time structures. This insisted that space-time can fold, warp, stretch, or generally be dynamical like an electric field. We are now ready for the master principle.

The General Principle of Invariance: *All systems of reference are equivalent with respect to formulation of the fundamental laws of physics.*

Einstein's principle of invariance proved so powerful in the discovery and underpinning of special relativity that he used it to unearth the hidden reality of space-time and the gravitational force, which we know as the general theory of relativity, but it would also prove to be more influential again. As we will see, this turn of the tides, of placing symmetries as a central principle, would also become the driving force to writing down the other nuclear forces.

Einstein arrived at his theory of general relativity by a thought experiment—his happiest thought—by realizing that the experience of an accelerated person in the absence of a gravitational field (like a person inside a rocket ship accelerating in outer space) is the same as a person standing at rest on a planet with a gravitational field. The converse is also true. If you are in free fall in the presence of Earth's gravitational field, your acceleration can "erase" the earth's gravitational field. In other words, if a person is isolated in an elevator and is in free fall toward Earth, they would feel no gravitational forces.[2] Einstein reasoned if acceleration could mimic the effect of a gravitational field, then gravity itself has to account for both acceleration and a gravitational field on the same footing. In other words, the law of gravity has to respect a symmetry between a constant gravitational field and accelerated motion without a gravitational field. How does one formulate such a symmetry? The strategy is to use what worked before, that is, special relativity.

General relativity requires invariance of the laws of physics regardless of the relative motion of observers. This requirement can be restated into a more general symmetry relating any type of coordinate transformation, called general covariance. For instance, a rotating observer will experience the laws of physics the same as a nonrotating observer.[3] The compromise for putting all observers on the same footing gives two radical features of gravitation. The first is that the gravitational force is not a force in the way we're used to thinking about it, but a feature of the curved space-time. In the Newtonian world an electric field could change the direction of a charged particle. In general relativity, particles will move in the straightest path possible. But in a curved surface, such as the surface of a sphere, the shortest distance between two points is not a straight line.

Imagine that you lived exactly at the equator of the earth and wanted to travel around the world using the shortest path to return home. The shortest path would be exactly a horizontal curve that circumscribes a great circle around the earth. This shortest path is not a straight line. Likewise, if space is curved then the shortest distance won't necessarily be a straight line in flat space. These straightest paths are called geodesics, and what we call the gravitational force is simply a geodesic motion in a curved space that a star, for example, bends.[4] The motion of Earth moving around the sun is nothing more than a geodesic due to the warping of space-time by the sun,

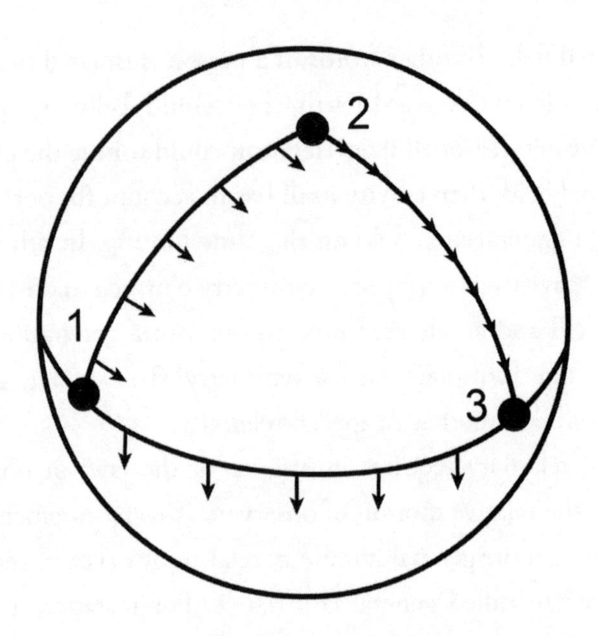

FIGURE 3: The intrinsic curvature on the surface of the sphere is measured by dragging and keeping parallel a flag around a closed circuit. The space is curved if the dragged flag fails to be aligned with its original orientation.

and our planet is just moving around the contours of the curved space-time.

To do physics we need dynamical equations for the objects of relevance. For electromagnetism, the electric and magnetic fields obey a set of four equations codified by James Clerk Maxwell. The principle of invariance allows us to repackage all four equations into one equation that is invariant under the coordinate transformations of special relativity.[5]

I suffer from having a bad memory and always forget one of Maxwell's four equations. So I was thrilled to realize that I can use the principle of invariance to write down one equation that contains all the other Maxwell equations.[6] And using the same approach, Einstein figured out that there is an object that can codify the dynamics of space-time that respects the general coordinate transformations.

Serendipitously this mathematical object was already discovered by mathematician Bernhard Riemann. The object, called the Riemann curvature tensor after its inventor, captures the dynamics of curved space-time in the presence of matter and energy. The famous Einstein equations of general relativity describe how the curvature of space-time is warped by matter and energy. And similar to the one Maxwell equation that contains others, the singular Einstein equation unpackages ten dynamical equations for the field of space-time.

Perhaps the most profound consequence of Einstein's principle of invariance is that space-time itself is not some empty, fixed stage but rather is a dynamical field. But this leaves an interesting question. Since the gravitational field is space-time, when the gravitational field vanishes, does space-time itself cease to exist? In other words, what is no-space? We will revisit this question, but answering it will require that we meet our other principles first. Let's jump into the quantum.

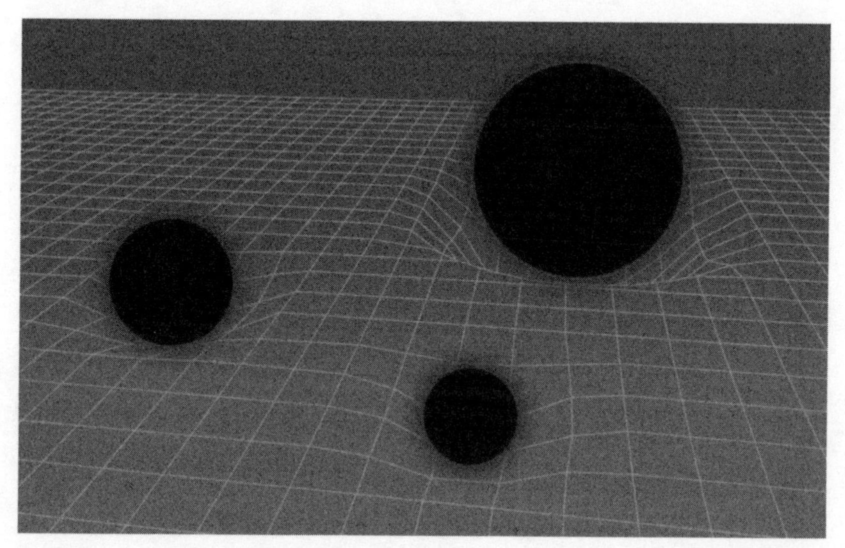

FIGURE 4: Three black holes affect the dynamics of space-time by warping the geometry of the space-time. This malleability of space-time has a range of physical effects including but not limited to gravitational forces and bending of light.

3

SUPERPOSITION

> If you think you understand quantum mechanics,
> you don't understand quantum mechanics.
>
> —Richard Feynman

It was sophomore year and the moment we physics majors had been waiting for. The last year and a half had been a tedious grind of problem sets that involved boring blocks sliding down inclined planes and systems of masses hanging on pulleys. But that day we sat attentively. It was our first quantum mechanics class, and we were waiting for Professor Lyle Roelofs, a tall man with a thick mustache in a plaid shirt, to deliver his first lecture. He began by saying, with a straight expression on his face, "I can assure you that after learning quantum mechanics, there will be no need to drop acid." I thought he was joking. Twenty-eight years later, however, I can report that he was right. Just when you think you understand it, when you stop to think of what quantum mechanics is saying about reality, there is really no need to drop acid—your mind has already experienced alternate reality.

Quantum physics underlies our electronic technology, and although we think about it as the physics of the very small, it is likely to apply to galaxies, too, as it probably describes the behavior of whatever particles make up dark matter. And, as we'll see, quantum mechanics is probably central to the existence of the universe, too. In fact, one of the reasons I became a physicist was Stephen Hawking and James Hartle's proposal of the wave function of the universe, which posits that the entire universe is a quantum mechanical system.

Perhaps it comes as no surprise that, just as no one ever described the general theory of relativity as being like an acid trip, quantum mechanics wouldn't offer up an obvious central principle like the principle of invariance. That doesn't mean we can't try to find one. If I had to choose one (which I guess I do!), it would be the principle of superposition of states.

Of all the strange things about quantum mechanics, I believe superposition is the main point of departure of the microscopic quantum world from the ordinary world of large things (like dogs and airplanes) of classical physics that we experience. Once we have this principle, it invokes strange features such as nonlocality, the uncertainty principle, and complementarity, which, if you're familiar with quantum mechanics, are all pretty strange, and if you're not, you're about to! These will form the basis of much of the cutting-edge mysteries we are currently wrestling with and exploring in these pages.

To appreciate the bizarre world that the principle of superposition of states describes, we should first understand the concepts of *superposition* and *state* independently. In classical mechanical systems, the physics of the macroworld you inhabit, every object has two attributes that describe what physicists call its physical state. One is its position, which is where the object is. The other is the object's momentum, which is defined by its mass and its velocity. This means that momentum is a vector, and includes both the speed of the object and the direction it is traveling in, as well as the object's mass. We can think about the physical state as a point in a two-dimensional space, where the y-axis denotes the momentum and the x-axis denotes the position. This space is called phase space. Phase spaces are key tools in analyzing all dynamical classical systems, whether a baseball or a spaceship—or even classical particles.[1] Each occupies a unique point in phase space. So if we have something simple, such as a ball in motion, we are able to predict with certainty what its state in the phase space will be in the future, once we apply to it the dynamical laws in classical physics. This is called determinism. A classical object can only be at a unique position and velocity at a given moment of time. I hope so far this makes good common sense.

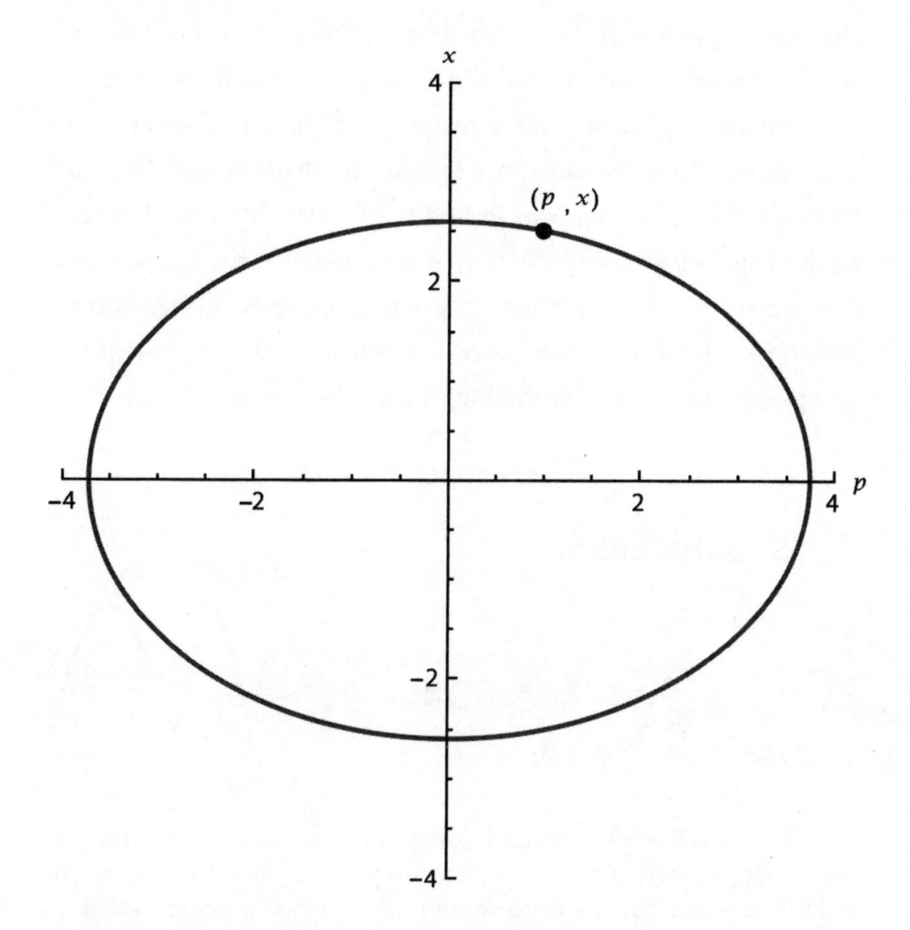

FIGURE 5: A phase-space diagram showing a particle's momentum and position.

You might think that, if states make good common sense, then it must be superposition that doesn't. But it can make common sense, too. In fact it is a common and useful property of waves. If we drop a stone on a calm pond, a periodic wave will propagate radially outward from where the stone hit the water. If we drop other stones nearby, other periodic waves will also propagate radially outward from those points as well, and sooner or later those waves will encounter each other and combine. This will result in a more complicated pattern of waves, where they can either add up or subtract from one another. The ability of waves to combine to make more interesting waves is at the heart of superposition: in fact, the word *superposition* literally means the process by which simple periodic waves of differing frequencies are added up to make a complicated-looking wave. An intricate-looking water wave is made up of billions of water molecules that collectively superpose to take the shape it has. This isn't just something that happens in bodies of water; it's why there can be dead spots if stereo speakers aren't set up correctly, and it is also how electronic music synthesizers work. Conversely, a complicated-looking wave can be decomposed—mathematically, anyway, if not in a pond—into a large collection of periodic simpler waves.

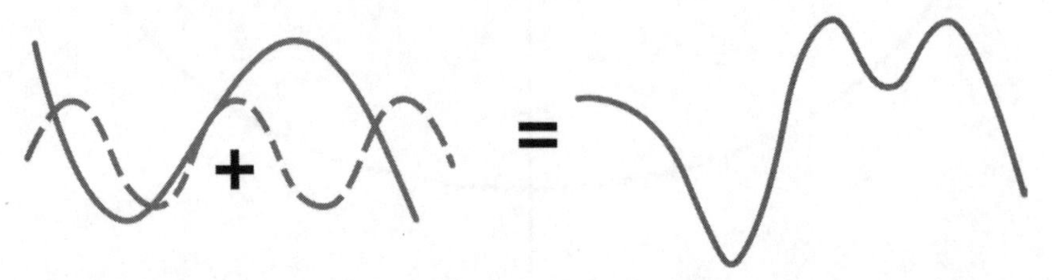

FIGURE 6: To the left, solid and dashed periodic waves can superpose (be added) to form a more complex wave to the right. Conversely, the right-hand wave can be decomposed into the two periodic waves on the left.

The ability of waves of water atoms to superpose made it all the more confusing for physicists when they first encountered the bizarre behavior of electrons in their experiments. In a nutshell, individual electrons were thought to be particles, but they exhibited wavelike behavior instead. Quantum mechanics was invented, in part, to deal with those experiments.

Here is the punch line: A single electron can exist as a superposition of many states at the same time. In other words, a single electron can be in many places, or have many momenta, at once. More generally, behavior of electrons or any quantum entity can be stated as follows:

The Quantum Superposition Principle: *A quantum system is expressed by adding many distinct states at the same time.*

A common example of quantum superposition concerns the position of an electron moving in space. The superposition principle says that the state of an electron's unique position in space is equivalent to the same electron having a wide array of velocities at the same time. The converse is also true. If the electron has a state describing it to have a given velocity, it can equally be expressed as a superposition of one electron having a very large range of positions.

You can see why Professor Roelofs thought this was crazier than tripping on acid: all your physical intuition about the classical world should go out the window. How can one electron be in several places at once? How can one particle be moving at many different velocities at the same time?

One of the most mysterious experimental displays of the quantum strangeness of superposition is the double slit experiment. Richard Feynman calls the double slit experiment "impossible, absolutely impossible to explain in any classical way, and has in it the heart of quantum mechanics." Here is how it goes: One electron at a time is shot toward a screen, which lights up where the electron lands. Between the electron gun and the screen is a barrier with two small holes that allow the electrons to pass through. At first, we see nothing out of the ordinary, just electrons landing and lighting up the screen like fireflies. After a large number of individual electrons arrive on the screen, however, they distribute themselves in a bizarre way. Since the electrons are particles, we expect them to pile up in two bundles directly behind the holes. However, the electrons land on the screen to create what's called a wave interference pattern, with areas with lots of light spots and areas with few. How did the individual electron particles know to avoid the location of the previous electrons to collectively exhibit a wave pattern later? What exactly is the electron doing when it arrives at the double slit? Is it behaving like water waves that go through both holes and interfere after they pass through?

We could partially explain the interference pattern if we assume that the electrons behave like a wave as it approaches the two nearby holes. The electron wave would split up and pass through both holes and recombine forming an interference pattern. Experimenters tried to find out if that's what was going on by putting detectors at one opening to see where the electron went through. What they saw instead was that the interference pattern on the screen completely disappeared. So, when we are not looking at the electrons during their flight, they behave like waves, and when we try to pin their positions down, they lose their wavelike properties and collapse into one location out of the many possible places they could have been. It's as if the electrons knew when they were being watched to change their wavelike properties. Ever since this bizarre situation presented itself, many physicists have attempted to reconcile it with either modifying quantum mechanics or giving it a new interpretation. There is still no general agreement!

Although experiments like the double slit experiment were and still are quite shocking, it turns out that most of our modern technology stems from the quantum mechanical understanding and ability to control the electron. My favorite is the solid state laser. Another is the nuclear magnetic resonance technology, which drives the MRI machine used to do brain scans in hospitals. Electrons carry a negative electric charge and have another property that physicists call half-integer spin. Spin is called that because, mathematically, it's something like the spinning of a top. Tops spin clockwise and counterclockwise; electrons have elementary spins we call up and down, and they behave like tiny magnetic north and south poles. Spins can be flipped by the application of a precise magnetic field, and this is at the heart of an MRI.[2]

The double slit experiment also inspired those in other disciplines, including both neuroscience and mysticism, with the behavior of the observed electron indicating that perhaps there is some undiscovered link between quantum mechanics and consciousness. These speculations continue to stir mystery, controversy, and confusion, even among Nobel Prize–winning physicists, and there may even be some truth to these intuitions. I will present my own suppositions later, but for now, it is important to see that this weird behavior stems from two ideas: that the electron's location in space is a superposition of position states, and when we make an observation we can only see the electron realize itself at one position.

This property is seen in a wide array of quantum mechanics and is currently being exploited to make quantum computers a reality. This conundrum is at the root of an issue in quantum mechanics called the measurement problem, a subject we must confront. The bizarre quantum superposition could also be a useful attribute for a superhero. Like many youngsters, I was hooked on Marvel comic books and imagined creating my own superhero. It turned out that it didn't matter that much that Stan Lee and I went to the same high school. Also, my drawing skills weren't that great, so I didn't end up pursuing that path. I am now amazed to see how quantum physics has entered in Marvel movies, such as *Ant-Man* and *The Avengers*.[3]

If I were to create a new character, I can think of a few properties of the quantum world that would fit the bill of major new attributes for a superhero. If my hero was fighting many villains at once, then he or she would have a major advantage over a nonquantum hero. A quantum superhero with the ability to be superposed in many positions at the same time could fight many villains at once. Of course, this raises some deep questions. If an electron causes trouble for us once we interact with it going through a slit, what's going to happen when the quantum hero interacts with classical villains? We will return to this issue when we discuss the measurement problem inherent in quantum mechanics.

But first it's useful to explore more quantum weirdness.

In 1923 a French prince named Louis de Broglie claimed that all matter possesses wavelike properties. First, by paradoxical fiat, de Broglie related the momentum of an electron to its wavelength, and further that all matter—anything with mass and momentum—also has a wavelength associated with it. This hypothesis was used to correctly explain one of the great mysteries of early twentieth-century physics, which is why negatively charged electrons don't go crashing into the positively charged nucleus of atoms. If we identify the wave as the electron's orbit around the nucleus, then the longest wavelength predicted by its mass would correspond to the closest orbit that it could make around the nucleus. The electron would not be allowed to go to lower orbits and fall into the nucleus because no lower wavelength orbitals would exist for the wave. This picture was also consistent with Niels Bohr's model of the atom, and why he had hypothesized that electrons existed in orbits that could only have discrete, integer values. This model correctly predicted and explained the absorption and emission spectrum of gasses that classical physics could not explain. Despite having a host of models, such as the Bohr atom and de Broglie's orbits, to account for the wavelike behavior of electrons, physicists still lacked a precise equation to describe what was going on. At least they did until Erwin Schrödinger returned from a short ski trip in the Swiss Alps with a most beautiful and elegant dynamical wave equation describing the wavelike superposition of quantum matter.

Schrödinger is one of my heroes in physics, and he inspired many young physicists of my generation as well. He wasn't the stereotypical "geeky" physicist that is often portrayed in movies and in TV shows like *The Big Bang Theory*. He had a host of other interests such as poetry, color theory, and Eastern mysticism, and I am convinced that his polymath tendencies influenced his creativity. While no one knows exactly how he came up with it, Schrödinger discovered a differential equation—which we call the Schrödinger equation—for all quantum systems. It is an intriguing fact that all our theories are described by differential equations; for some reason, after all the elegant formulations of a physical theory, when the rubber hits the road, physicists end up solving differential equations. The solution of the Schrödinger differential equation is a mathematical function that contains all information, including superposition of states, about quantum systems: the wave function ψ.[4]

The Schrödinger equation is so aesthetically beautiful, I can't help but write it down.

$$\left[-\frac{\hbar^2}{2m}\nabla^2 + V\right]\Psi = i\hbar\frac{\partial\Psi}{\partial t}$$

This singular equation determines the behavior of the wave function, denoted by the variable, spoken as *psi* (and sounds like *sigh*). It describes how the wave function of a quantum system interacts, and changes in time. This equation predicts all quantum phenomena, including the periodic table, physics of a neutron star, and even the semiconductor device in your smartphone. The equation simply says that the temporal change, or evolution, of a quantum wave function (right side) depends on the energy, which encodes interactions acting on it (left side). Once we know the evolution of the wave function, we get new insights into and predictions for the quantum system at hand. Some of these properties led to the invention of gadgets like transistors, the heart of all computer-driven technology. The superposition principle emerges naturally from it. This is because the Schrödinger equation is what is known as linear. This means that if one finds two independent solutions of the wave function, then the sum of those solutions would also be a solution. Periodic waves obey this mathematical property of linearity. Thus, the wave function describes waves that can be decomposed into an infinite set of simple harmonic waves of differing frequencies.

So what is the electron doing, according to the Schrödinger equation? Schrödinger interpreted the electron as a highly peaked material wave, like a pulse: the wave is the electron. But with this interpretation a natural paradox presents itself: If a highly peaked wave function is to describe a single electron, what does it mean when the electron wave function hits a barrier and splits in two? One common feature of a moving wave is that if it starts off highly peaked, at some later time it will spread. But what becomes of the particle when the wave loses its shape and splits into many pieces? Are we to think that the electron will split up into many electrons? Precision experiments from particle accelerators reveal that the electron is an elementary particle and is not a composite of any smaller unit of matter. So if the wave splits, the electron certainly does not split in two. It could be that the electron does lose its identity when it encounters an environment that enables it to shape-shift. We will return to this issue. For now, let's accept the electron as an indivisible and independent unit.

Schrödinger's interpretation that the wave function represents a material wave whose shape describes the position of a particle was problematic, and it forced physicists at the time to seek an interpretation of the wave function to also make sense out of a quantum particle being in many states at once, as is suggested by the double slit experiment. Max Born, another leading quantum physicist, provided an ingenious interpretation of the Schrödinger equation that dealt with the problem of a splitting wave: he argued that the wave function is not a material wave but a wave of possibilities, so that where the wave has the largest amplitude the particle is more likely to be there, and correspondingly less likely to be where the wave's amplitude is lower. Applying the Born interpretation to the double slit experiment, the wave pattern we see is the result of a probability wave that goes through both slits. When we look to see which hole the electron is at, the probability superposition collapses into one position outcome and the interference pattern disappears. This still doesn't solve every problem with the double slit experiment. You should scratch your head and wonder how the act of observation collapses this probability distribution in the wave function to the observed value. Also, what does it mean for a probability wave to go through the slit?

The competing interpretations of the wave function forced the architects of quantum mechanics to take philosophical stances about how to interpret quantum mechanics generally. Albert Einstein, one of the founders of the theory, took a realist stance. Realism demands that there is an objective physical world that is independent of our existence. This seems reasonable since the universe had existed billions of years before stars, planets, and humans came on the scene. Saying that a probability wave goes through a hole avoids saying exactly what the electron is actually doing in that region of space. In accounting for the double slit experiment, a realist posits that the quantum theory should explain what the electron does when it encounters the two open slits and the resulting interference pattern, with or without the presence of a measuring interaction.

There is a realist interpretation that does this, the de Broglie–Bohm pilot wave. In the de Broglie–Bohm theory there is no superposition in the wave function, but a particle and an invisible wave that propagates through the double slit. This electron surfs the contours of the wave and ends up tracing out the interference pattern of the wave. This de Broglie–Bohm interpretation still uses the Schrödinger equation but trades off the probability weirdness for another kind of weirdness, which allows for nonlocal interactions between physical objects. In the de Broglie–Bohm theory, the wave can cause faster-than-light correlations between different electrons. If the wave changes in one region, an electron motion will be instantaneously affected at a distant region.

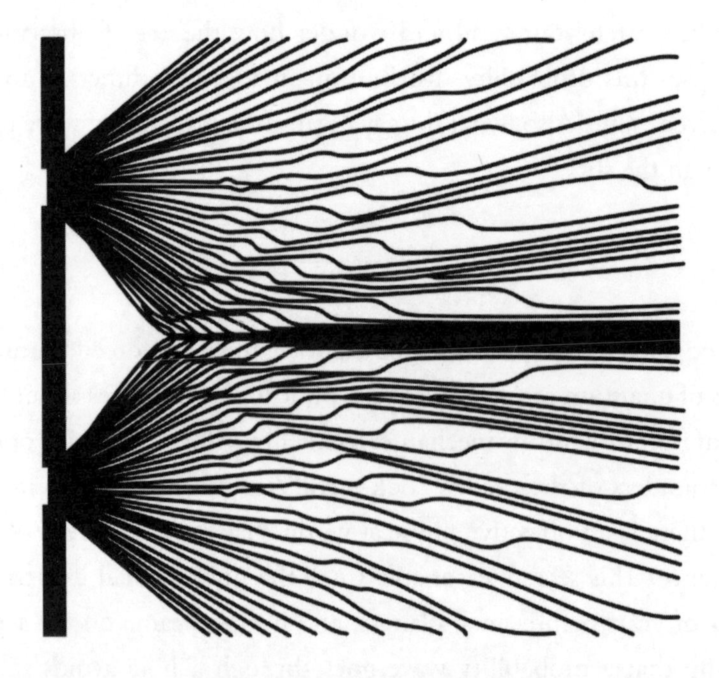

FIGURE 7: The double slit experiment as represented by the de Broglie–Bohm pilot wave theory. The lines that emanate from the two slits are the trajectories of the electrons along the contour of the nonlocal quantum potential surface.

Niels Bohr and his followers rejected realism and actually formulated quantum mechanics, the form that is taught in most textbooks, to avoid asking questions about the electron's reality when it's not being observed. Some such as Werner Heisenberg and Born even questioned the existence of particles until an interaction takes place. Influenced by Eastern philosophy such as Buddhism and Taoism, Bohr believed that our classical experience and perceptions were incompatible with the intrinsic duality that quantum entities possessed. According to Bohr, quantum entities possess pairs of contradictory qualities, such as wave and particle, energy and time, electric and magnetic, spin up and spin down. In Taoism the yin and yang represent contradictory or opposing qualities that coexist to describe the whole. Likewise, the whole electron possesses both the contradictory wave and particle properties at the same time. It is through our macroscopic measuring devices that we see either the particle or the wave properties revealed, and not both. Bohr elevated this duality to a principle he called complementarity.

FIGURE 8: A visual representation of complementarity. There appears to be two independent images. Upon closer inspection, the boundaries of both images define each other.

Some of Bohr's brilliant young followers incorporated his complementarity principle into the actual mathematical formalism of quantum theory. The Nobel Prize was awarded to a young Werner Heisenberg, who formulated the uncertainty principle, which is rooted in complementarity. To see this a bit more clearly let's combine Bohr's wave-particle complementarity with the superposition principle to see how uncertainty follows naturally. According to the superposition principle, simple periodic waves of different frequencies can be added up (superposed) to give a more complicated wave. I can add up many periodic waves of differing frequencies to give a pulse, which approximates a particle like the electron. A quantum state described by a wave with definite frequency is actually the same as a particle with a definite velocity. This happens because the speed of a periodic wave is proportional to how rapidly it oscillates—its frequency. And according to the superposition principle it takes a very large superposition of many different velocity states to approximate a state of definite position of a particle.

This feature of duality and uncertainty marks a profound difference between the classical and quantum worlds. In classical systems, the dynamics are given in pairs of physical quantities, for example, as we discussed position and momentum. These dual pairs come together to form phase space, and in principle we can know everything about the position and momentum at the same time with complete accuracy. On the other hand, in quantum mechanics we can only know one of the quantities with precision, at the expense of randomizing the other. The key idea is that dual qualities like position and momentum simply and inevitably disrupt each other's certainty.

Heisenberg came up with a thought experiment, the Heisenberg microscope, to highlight this compromise. He imagined that if we observe an electron, light has to bounce off it, exchanging a definite amount of momentum, which will change the electron's trajectory, hence randomizing its subsequent position. Consider a state with a unique definite frequency. I ask you, where is the wave? If I pick a point on the wave, it looks exactly the same as any other point if I shift along by one complete cycle, so any point on the wave is not a unique position. Therefore, the specific location of the wave is indeterminate. A perfectly periodic wave is everywhere! Now imagine a pulse, similar to what we see on the devices that measure your heartbeat. The position of the pulse is located where the pulse is highest in height. A pulse is a wave that is very localized in space. However, the frequency of a pulse is indeterminate because a frequency is defined to be a quantity that depicts how often a wave repeats itself in a given amount of time. The singular pulse only repeats itself once. So, a periodic wave has a definite frequency but an indefinite position. And a pulse (now think of it as a particle) has a well-defined position but an indefinite frequency. There is a tradeoff. Quantum states that are localized in space are indefinite in their momentum, and those with precise momentum have delocalized in space.

Einstein was vehemently against the idea that quantum mechanics, and so nature, was fundamentally operating on chance. He famously said, "God does not play dice," reflecting his realist stance on physics. Bohr responded, "Don't tell God what to do." Despite the witticisms and counterarguments, Einstein persisted in disagreeing with Bohr and his followers and set out to find the death blow to the anti-realism of quantum mechanics. Einstein teamed up with his own duo of young researchers, Boris Podolsky and Nathan Rosen. Together, they wrote a paper containing what is known as the EPR paradox. EPR showed that special states exist in quantum mechanics such that two particles can have opposing features; these are called entangled states. A version of this thought experiment involved a pair of photons that is created by the decay of a particle at rest. If the photons start out in a spin-zero state and fly off in opposite directions at astronomical distance, the measurement of the spin of one photon will predict with 100 percent probability that the other spin has to be opposite, because the photons are in a spin-zero entangled state. The other spin will be communicated instantaneously. EPR coined this as "spooky action at a distance," and it was seen to be in opposition to the fact—a critical feature of Einstein's special theory of relativity—that no signal can travel faster than the speed of light.

Ironically for Einstein, what was intended to be a potential death blow to the quantum turned out to be a brilliant prediction when it was experimentally observed. These experiments revealed that quantum particles exhibit nonlocal effects. For example, measuring the polarization of a photon instantaneously determines the spin of its entangled pair. Simply put, nature is nonlocal. The other question was whether the spin was real before it was measured—that is, whether it was determined by what is called a hidden variable of the system, rather than being a random result of the process of measurement, as Bohr and his followers believed. According to the groundbreaking proof of Irish physicist John Bell, this nonlocality in quantum mechanics has to be the result of the action of a hidden variable. Interestingly, Bell used the de Broglie–Bohm interpretation, based on the yet-to-be-observed pilot waves, as an example of a nonlocal hidden variable theory.

All of this points to a fundamental tension in quantum mechanics. At one level quantum mechanics is completely deterministic. According to the Schrödinger equation, if one knows the initial state of the wave function at some initial time, then one knows exactly what the state will be at a later time. However, because the wave function represents a superposition of possible outcomes—it's a probability distribution—then if a measurement is made, we will see only one of such outcomes. In other words, an observation collapses the function in an indeterministic way. This tension was rigorously investigated by John von Neumann, one of the great mathematicians of the century. Von Neumann, along with Eugene Wigner and Wolfgang Pauli, argued that the measurement problem required a resolution that would go beyond the current formulation of the theory. They even went so far as to argue that consciousness itself played a role in collapsing the wave function. To the contrary, Bohr and his followers asserted that there was no problem based on complementarity, and they embraced the inherent contradiction in quantum mechanics. According to Bohr, there is a sharp divide between the quantum world and classical measuring device, and the instantaneous collapse of the wave function is a feature of this divide. Decades of Nobel Prize–winning physicists have landed on opposing sides of the measurement problem.

Despite these ongoing and unresolved issues in quantum mechanics, its formulation as a probabilistic theory where superpositions can collapse upon measurement has passed a century of precision experiments and applications. However, as we shall see, these issues will come back to haunt us when we connect quantum mechanics to gravity and the early universe.

We'll get back to that. After all, we've still got one more physical principle to go. But before we do, let's explore what happens when we attempt to merge the invariance with the quantum principle.

4

THE ZEN OF QUANTUM FIELDS

Most mornings as a kid I would hear about my mother's experiences as a night-shift nurse at Montefiore Hospital in the Bronx. While she was getting me and my brothers ready for school, she would tell us about some of her patients and their various medical predicaments. These stories ignited big questions about my own mortality, about the fact that my time on this planet was finite and even uncertain. In hindsight, this was a driving force behind my decision to study physics, as the field provided a rational approach to understanding the physical world and our place in it, and would help make sense of reality and the big questions I had about existence.

As the years went by, my questions about what physics should say about our place in the universe got overshadowed by the rat race of publishing articles and landing a permanent faculty position. I didn't forget about them, however. In recent years, I have come to the conclusion that, at least metaphorically, modern physics is starting to make contact in a non-woo-woo way to some tenets of Eastern philosophy. Now I can see why Schrödinger and Bohr read that stuff. Some of these connections have been made in other works, such as *The Tao of Physics*. My goal here is to add what I consider to be a sound new metaphor to this mix of connections, based on the fusion of invariance and superposition, that is relevant to the question: Why is there something rather than nothing?

My first introduction to Eastern philosophy was a book entitled *Zen Mind, Beginner's Mind* by Zen master Shunryu Suzuki. Suzuki had a most interesting metaphor for existence. Imagine a stream flowing toward a waterfall. When the water leaves the cliff a droplet of water leaves the stream and at some point, rejoins the stream. Life is like that droplet of water and the stream is like the universe. During the time between when we are "born" and when we "pass away," we are like that water droplet, feeling like we are separate from the universe as a whole. But before we are born and after we die, we are a part of the stream, the universe. I remember wanting there to be some truth to Suzuki's metaphor. So, in what way could we be connected with the universe in the poetic manner described by Suzuki? According to Zen philosophy this merging can be only experienced. But I have never experienced Suzuki's claim, since I experience having a separate and localized body that's made up of matter as I occupy and move through empty space.[1] So in place of experience theoretical physics will have to suffice.

There was a Zen monastery twenty minutes from Brown University, where I was a graduate student: the Providence Zen Center. My friend and physics classmate Claire was a regular Zen practitioner and brought me and a few friends along to engage in a morning of formal practice. The morning included, among other forms of meditation, my favorite, breakfast meditation—when eating, just eat. There were others: after fifteen minutes of chanting words that I didn't understand, we sat in silence for three sets of half-hour sitting meditations followed by walking meditation. This cycle of sitting and walking meditation would continue for a few hours. Then the Zen master came out and gave a little talk. She said, "You, me, this table, this universe are all made from the same fundamental substance . . .

form is emptiness and emptiness is form." I later learned that many Zen practitioners seek to attain a mental state called satori, where they can experience being one with this fundamental substance.

How amazing that at the same time I was learning the mother language of physics called quantum field theory, and the modern view of physics confirms experimentally what these masters have experienced subjectively about who we are—necessarily connected to the universe. But if this insight is true, it must be linked to the matter that we are made of, atoms and their associated force carriers. And we will now see that our popular and pedagogical view of matter has been flawed. To make this clear, let us first look at one of the fundamental building blocks of matter, the electron.

One of the biggest misconceptions that is nevertheless routinely taught is that things are made of elementary particles that form the building blocks of atoms and molecules. It is the case that from a historical point of view, atoms and a zoo of elementary particles were discovered first, but when physicists were trying to reconcile the quantum mechanical nature of these particles with special relativity, the particle picture was overthrown by a deeper reality of the quantum field. This field picture of all matter came as a necessity of trying to reconcile the physics of an electron moving close to the speed of light in an atom. To properly understand the electron's behavior in this relativistic context, we must find a way to merge special relativity with the electron's controller, quantum mechanics. When you do attempt to merge the quantum with relativity, you immediately see, according to the invariance principle, that the basic equation of quantum mechanics is not invariant under the space-time symmetries of special relativity. The main reason is that ordinary quantum mechanics gives time a preferred status over space, and in relativity, they are on the same footing. In 1929 English physicist Paul Dirac ingeniously found a way to unite special relativity with quantum mechanics.

In hindsight, the hint of this unification comes from electromagnetism, a classical field theory that Einstein showed is already invariant under special relativity. The idea of the field was first introduced by English scientist Michael Faraday to explain his ingenious experiments involving moving magnets, which he found would create electric currents in nearby circuits. To explain the action at a distance between the magnet and the current-carrying wire, Faraday stated, "I believe that magnetism is actually propagating itself through this invisible field of influence." Faraday also said, "I believe that Electricity has this invisible field of influence and so does gravity." His peers rejected the claim about fields as idiotic. This "invisible field" was considered heretical because the paradigm of the time was that of a mechanical universe. Similar types of heresy today would be considered "woo-woo" by critics on YouTube channels. Biographers say that Faraday died of heartbreak because his field idea was not accepted during his lifetime. Ironically, long after Faraday's rejection from the scientific community, schoolchildren still play with magnetic filings that trace out magnetic field lines emanating between the north and south poles of a magnet.

If you really stop to think of invisible agents acting to impart forces on objects through empty space, the field concept is actually eerie, the stuff of witchcraft. It was not until later in the century that the field idea became accepted. And as we will now see, the field concept will become a central paradigm underlying modern physics.

James Clerk Maxwell unified a disparate set of electric and magnetic properties into a coherent framework grounded in the reality of an electromagnetic field. We can think of the electromagnetic field as a continuous substance that is distributed throughout space-time. Electric and magnetic charges will bend the field in definite ways.

Conversely, a warped field can exert forces on a charged particle. Disturbances of the electromagnetic field can create waves that move at the speed of light, commonly known as electromagnetic waves. When these waves vibrate at a frequency on the order of a trillion cycles per second, we perceive it as visible light. Einstein earned the Nobel Prize showing that waves of light can also behave like a quantum particle, the photon, and the wave-particle duality had to be reckoned with. It took Dirac, German physicist Pascual Jordan, and Born to realize that the fundamental substance of the photon is the electromagnetic field that is emitted when the electron makes quantum jumps from an excited state. In this case, harmonic vibrations, or quanta, of the electromagnetic fields gave rise to the photon. If the photon is a particle excitation of the electromagnetic field, what about the other particles of nature—are they related to a field?

What Paul Dirac discovered was that to unite special relativity with quantum mechanics, the invariance principle with the quantum principle, something radical about the nature of the electron had to be compromised. He finally figured out, in the spirit of Einstein, how to make the Schrödinger equation invariant under the symmetries of special relativity. After succeeding, he found a new symmetry relating the electron with a mirror electron with negative energy. In physics when we encounter situations with negative energy, we run for cover. Negative energy is like falling down a hill with nothing to stop the fall. But it gets worse than that because all of that energy will cascade through quantum interactions to the electromagnetic field, creating an unfathomable explosion that would destroy an entire galactic system. In Dirac's case, he used symmetry to ingeniously reinterpret and repackage this negative energy electron as a new particle with positive charge and positive energy and called it a positron, the antiparticle of the electron—a new particle was born! Two years later the Nobel Prize–winning discovery of the positron was made by Carl Anderson, confirming quantum field theory. With the reality of the positron this meant that when an electron and a positron interact, they annihilate each other and their masses convert to energy to produce a photon, according to Einstein's famous equation that states that matter is equal to energy. Or an electron and positron can spontaneously be created from a photon with an energy twice as much as the rest mass of the electron and positron. This led Dirac and others to no longer think of the electron particle as a fundamental entity, but a part of one underlying electron-positron field.

A faithful picture of a quantum field is to imagine the field as the height of a blanket supported by oscillators (springs). At different points on the surface of the blanket the spring will vibrate differently. Now imagine that all these springs are connected to each other. Then the vibrational pattern of the blanket could be quite complicated. There are special vibrations of the springs that can oscillate in sync with each other. Let's carry the analogy further and imagine that there is a mass on each spring. Straightforward spring mechanics tells us that the rate of vibration of the spring, its frequency, is proportional to the mass. The heavier the spring, the faster the spring bobs back and forth. So, when the field vibrates at the frequency equivalent to the mass of the electron, the electron can be created as a particle from the field's sympathetic vibration, or resonance. The particle's creation from its underlying field is analogous to Zen master Suzuki's water droplet that emerges from the cosmic ocean.[2] Likewise, particles are transient entities that could emerge and return back to their mother field. Where do these fields live? The electron field is omnipresent and like the electromagnetic field occupies space-time of the universe. De Broglie's hypothesis, which we discussed, is realized because the vibrational standing wave pattern of the electron's field can manifest itself as an electron particle: the particle is nothing but quanta of the electron's field vibration.

This leads to a new question about the rest of the matter around us—could it all be emerging from an excited field? We have a hint

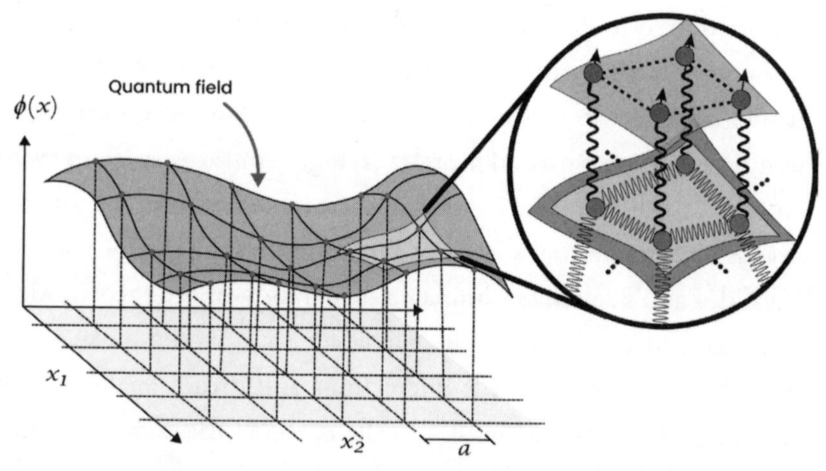

FIGURE 9: The depiction of the quantum field as a collection of oscillators that can resonate to create quanta of particles from the field's vacuum state.

from de Broglie's hypothesis that relates the mass of a quantum particle to its wavelength. Amazingly, all matter fields including the electron share exactly the same building blocks in that they are all fermionic fields, named after inventor Enrico Fermi. Fermions, the basic building blocks for all matter, include quarks—the building blocks of protons and neutrons—neutrinos, and other heavier and more exotic particles. They all have the key property that they refuse to occupy the same quantum state. This property is called the exclusion principle and was discovered by Wolfgang Pauli, who also discovered the particle called the neutrino. Fermion states are given by their position and spin. So, if a fermion is at a given position with, say, spin up, then another fermion with spin up can never occupy the same position. But because of the exclusion principle, if we continue to pile up fermions due to their exclusionary behavior, they will form atoms, molecules, and the various macroscopic forms of matter in our world.

Here is a story of blackness in physics. The book *Atom and Archetype*, which Chris Isham had introduced me to, includes letters exchanged by Pauli and Jung that reveal that the exclusion principle presented itself to Pauli in a series of dreams. Pauli had a reputation for being a harsh critic of other theorists and would swiftly catch mathematical errors and inconsistencies. He is famously known for humiliating those with bad ideas by saying, "Your theory is not even wrong." In other words, a theory can be incorrect, but at least the process and intellectual sharpness in developing the theory can still be valid. Pauli's statement implies a denigration of the theorist's intelligence and process as being laughable. So, it would make some sense that he kept secret his interactions and dream analysis with Carl Jung. It's especially ironic that the kernel of the exclusion principle came from dreams. Would he have been taken less seriously or shunned by other colleagues if they knew this?

So, if all matter fields are fermions, what gives them their distinct flavor? In other words, what makes an electron different from a quark or a neutrino, given that they are all fermions? The quick answer is that fermion species can carry different types of charges (electric charge being one familiar example). We should return to a quick discussion of fermion field and see what it says about these differences. Let's consider the electron field, whose electric charge sources an electric field that fills space. How does the electric charge relate to its parent fermion field? It so happens that the electric charge is beautifully encoded in the fermion field as a special phase of the field's vibration. The important idea of phase is central to both wave and field properties. Imagine two boats rocking up and down due to the surface-of-ocean waves. If both boats were bobbing at the same frequency and their high and low points matched, they are said to be in phase. On the other hand, two identical waves can be out of phase if their positions do not exactly match up. Here's an interesting fact: the phase of the electron field controls the value of its electric charge. Now there is no reason for the phase of an electron field at different locations to exactly match, much as rocking boats at different parts of a tumultuous sea would only be in phase by an unlikely coincidence. Nevertheless, experiments have revealed that the value of the electron's charge is the same regardless of the location, on Earth, the moon, or anywhere. This means that the electron field has the same phase, like a couple of miraculous boats. But what prevents a physics demigod from making the phase of the electron in New York and Trinidad different, so that Trinidadians get more electric charge for their dollar? A special form of the principle of invariance comes to the rescue—this is famously known as local phase invariance.

Quantum mechanics does allow our demigod to mess with the phases of fields at different locations, and this will problematize the fact that we see the same charges everywhere. But it so happens that there is a magical way to undo the damage the demigod can make to the phases. If we introduce a photon field to interact with the electron field the right way, then every time the demigod changes the phase of the electron, the photon field also makes a compensatory change in its phase so as to erase the change in the would-be difference of the electron's charge. And this magically happens if the photon field interacts with the electron in a simple way.[3] This type of invariance was invented by the genius Hermann Weyl, and he called it phase invariance, or more commonly gauge invariance or its German origin *Eichinvarianz*. Gauge invariance requires that the electron field interact with light and guarantees the universality of the electron's charge!

Now we can understand what makes the fermion matter fields all different from each other—it's their charges, or phase symmetries, which turn out to be the organizing principle of the standard model of particle interactions. Once I tell you that the electron has a phase invariance, you can immediately write down the full electromagnetic theory, with the unique interaction between the electron and the photon. And it turns out, the same reasoning goes into determining the other two nuclear forces, the weak and strong forces, as they are simply applications of phase symmetries applied to electromagnetism. The phases of the weak and strong interactions have more symmetries and give more involved interactions between the force-carrying particles and the matter fields that carry the weak and strong charges with which the force-carrying particles interact, but all these forces strictly come from Einstein's principle of invariance, this time applied to the phases of the fermion fields.

Consider electrons in a star millions of light-years away from Earth. While there are many electron particles, there is only one electron field that they are born from. Those electrons, and the electrons that are in you and me, were created as quantum excitations from the same universal electron field in the early universe. In fact, every other particle, including quarks and neutrinos, in the universe are quanta that emerge from their corresponding field vibrations. It's in this Suzuki sense that we are tethered to the fundamental fabric of the universe. All the particles that comprise us are quantum vibrations of the same quantum fields that extend across the universe. But does it end here? What is the relationship between space-time and the quantum fields that permeate it? What incited these quantum fields to generate the particles that occupy our universe to turn into stars, planets, and us? And why, since Dirac demonstrated that antimatter must have been created pairwise with all that matter, do we not see very much of it around?

Underneath the answers to those questions lies our third principle: emergence.

5

EMERGENCE

A handful of elements from the periodic table come together to create living things like you and me. Yet, the elements themselves are lifeless. How does life emerge from these building blocks? This question is at the heart of emergence. The first two principles we discussed, invariance and superposition, combined to give us the quantum fields we recognize as the building blocks of matter and subatomic forces. Following Paul Dirac's prediction of the positron, new symmetries in nature were discovered as shorter distance scales, and these new symmetries were experimentally probed with high-energy particle colliders. And these symmetries also functioned to specify the nature of the interactions that fields exhibited with other fields and their associated zoo of particles.

Throughout the last century, physics has been dominated by the quest to identify the natures of those subatomic forces and a theory of everything that unifies them. The idea has been that one day we would be able to find all the fundamental building blocks of the universe, as well as the rules for their interactions. This approach is referred to as reductionism, and is essentially motivated by trying to figure out how to build the universe from the ground up. Superstring theory is an example of such a unified theory. In string theory, the basic building blocks are one-dimensional strings. However, even if we were able to discover such a unified theory, it isn't clear that it would be able to explain phenomena such as life or consciousness. To the contrary: there's a very good chance that the fundamental theory could not arise from a reductionist approach.

So we are at a crossroads. On one hand, a major program of twentieth-century physics took us down the successful road of reductionism. The symmetries that were discovered informed us about the patterns and relationships between the elementary forms of matter and the precise ways that they interacted with each other. Indeed, the notion that what is fundamental became synonymous with unveiling more symmetries, at least to some leaders in physics, and that perspective played a prominent role in my own career. I was just starting my dissertation work using the exotic symmetries in superstring theory to solve some of the problems in early universe cosmology. We know from astrophysical observations that the universe is expanding. At the earliest stages of the universe's history, its environment was exceedingly energetic, hot and dense—conditions not at all unlike those at a collision of particles in a collider like the Large Hadron Collider (LHC). So, from the patterns found in collider experiments, we expect that these symmetries generated by superstrings were activated in the very early universe, where superstrings are expected to be the key players. My thesis exported a special symmetry from string theory into cosmology, called target space duality, or T-duality, which treats the physics in a large region of space and a small region of space as being the same. So, as we approach the big bang singularity, where the universe approached microscopic distances, we could avoid the big bang singularity with T-duality. The dream was to use cosmology to test unified theories like string theory, or other approaches to quantum gravity. Superstring cosmology is still an important topic in cosmology, and I still devote some of my research efforts in this direction. So far, the expectation to unveil more symmetries at the shortest distance scales works theoretically.

On the other hand, there were clear failures of the reductionist regime in particle physics. One of the most famous examples of emergence in quantum physics is superconductivity—really, it's the poster child of emergence in physics. In 1911, Nobel laureate Heike Kamerlingh Onnes observed that when he lowered the temperature of a metal close to absolute zero, the electric current would flow with zero resistance. There was no reason to expect how and why the billions of electrons, which repel each other as well as experiencing impurities in the metal, should superconduct. After all, lowering the temperature does not seem to prevent the electrons from repelling each other or get rid of the impurities in the metal. Many great physicists, like Einstein, Schrödinger, Heisenberg, Lev Landau, and Feynman, worked on superconductivity. During this time, many thought that completely new physics was needed, perhaps a new law, to account for superconductivity. And it took forty-six years for the trio of John Bardeen, Leon Cooper, and John Robert Schrieffer (developers of the "BCS" theory) to show that good old quantum mechanics and electromagnetism were enough. Superconductivity doesn't supplant them. It emerges from them.

The Principle of Emergence: *Systems with interacting elementary constituents can exhibit novel properties that are not possessed by the constituents themselves.*

The reason I am promoting the phenomenon of emergence to a principle is based on Einstein's criteria for a principle, because, to borrow his words, "[A principle is realized by] perceiving in comprehensive complexes of empirical facts certain general features which permit of precise formulation." What are the complexes of empirical facts in emergence that can transcend its context? There are many examples of disparate and seemingly unrelated physical and biological systems where we see emergence. Emergent behavior also happens in populations of living organisms. Groups of ants can collectively build a bridge of ants so that others can cross a barrier of water. The origin of life itself is argued to be an emergent phenomena. A unicellular bacteria has autonomous properties, such as motility, replication, and metabolism that its individual molecules, like proteins, do not exhibit.

But what is at the heart of emergence? How does a system "know" to exhibit novel collective behavior? These are hard questions that are currently being researched, and there are some partial answers. A universal aspect of emergence is the relationships between the emergent system and the parts that make it up. Although the emergent property is novel relative to its constituents, there is an interdependence between the emergence and the constituents. For example, the system of atoms that gives rise to an emergent liquid property depends on the collective behavior of the individual atoms. However, in the solid state, the atoms are on average located in a repeating array, forming a large-scale crystal. A simple example of emergence can be seen right in front of your face.

The renowned theoretical physicist Nigel Goldenfeld, who now directs an institute that focuses on finding links between emergence in physics and biology, has an experiment that we can all do to demonstrate an emergent phenomenon. Here is how it goes. First, push your hand forward in empty space. Second, get a chair and push the chair until it falls on the floor. The fact that the chair moved and fell on the floor is emergence at work.

This might sound strange, but it's true. When you moved your hand, it was interacting with the air, which is actually a fluid made up of air molecules. The chair, on the other hand, is a solid. The atoms of the fluid and the solid obey the same laws of atomic interactions, yet, despite the sameness of their atomic interactions, the solid state has new emergent laws of physics. That is, it has new long-range forces, such as a rigid elastic response from your hand pushing against it. The origin of these new forces in the solid state arises from the statistical properties, or collective behavior, of the billions of atoms. If we work at the level of the description of the solid, we can't deduce what the underlying interactions of the atoms are. At the level of the solid state, all we can deduce is that the continuum description, rigidity, elasticity, and so on emerges from the collective behavior of the atoms.

The same goes for superconductivity. What I find interesting in that story is how BCS hacked superconductivity, not just because it's interesting science, but because it gives some guidance for current problems that we are trying to solve. After all, some of the problems that we consider to be impossible to solve have been around for a shorter period than superconductivity.

I had the good fortune to hear some of the story of superconductivity directly from Leon Cooper, who was my first PhD adviser before I changed fields to quantum cosmology. During my first year of graduate school, I didn't know who Cooper was, and no one told me that he had won the Nobel Prize. He had the flair of a Shakespearean actor, wore fine Italian suits, and sported shades on a well-groomed head of jet black, wavy hair. During our weekly departmentwide talk, where famous researchers would present their results, Cooper would sit at the front of the room and ask what seemed to be naive questions, the type a schoolchild would ask. And this was exactly the quality of mind that, among other things, enabled him to access the ingenuous insight that would crack the nearly fifty-year-old problem of superconductivity.

Superconductivity was couched in the subfield of the physics of solids, known as solid-state physics.[1] One of the leaders of the field, John Bardeen, who previously shared a Nobel Prize for the discovery of the transistor, had been tirelessly working on superconductivity for years with no luck. Bardeen was well aware of the decades of failed attempts to explain superconductivity, and he decided he had to get an outsider's perspective and tools to bring new life to the problem. So, he sought out a theorist that had a fresh pair of eyes and wasn't jaded by the biases that may be formed by experts in the field.

I remember the serendipitous early summer day I was driving from Hanover, New Hampshire, to New York City and decided to stop by Brown with the hopes of seeing Professor Cooper. It had been almost a decade since I had seen him in person. I caught wind that he was retiring (but still fully engaged in research, as he still is). During summers, many faculty are traveling to conferences, but when I got to the physics building at Brown, Cooper was there in his spacious office, filled with books and covered with blackboards, doing a calculation. We sat down for a chat. He immediately asked me what I was working on.

I thought that I was going to impress him with my new take on the matter-antimatter asymmetry in the universe. Cooper stopped me and said, "You should find a real problem and solve it. Many people put their hands up in surrender when a problem gets too hard and claim it's impossible." I took this to be both a challenge and validating. What might at face value look like Cooper's rejection of my idea I took instead as him holding me to a higher standard and the expectation to solve a big problem. Even today, I try to live up to my former adviser's version of tough love with my students—to recognize and help awaken their hidden talents. Up till that point, I was playing it safe and avoiding physics problems that I thought only the most able of physicists should have the permission to work on. I asked Cooper, "How did you solve superconductivity?" What he said gave me some strategies for approaching my own problems.

Cooper went on to tell me that he was trained in another field, theoretical particle physics, and had mastery of the techniques new to that field, such as Feynman diagrams. When he was invited to work with John Bardeen in condensed-matter physics, as the field working on superconductivity was known, he had an unsullied and less-biased take on the nature of the problems those physicists were facing. For one, as we've seen in this book, particle physics concerned itself with discovering the nature of subatomic forces by exploiting the quantum scattering processes between elementary and nuclear particles. Solid state environments concern the behavior of billions of interacting electrons in an environment filled with other atoms usually organized in the form of a periodic crystal lattice. Superconductivity was both a conceptual and mathematically technical dragon to slay. One major obstacle was that the problem seemed to require solving the Schrödinger equation for a wave function of billions of electrons interacting with a lattice of metallic atoms—the many-body wave function. No one working on the problem, no matter how technically skilled, has been able to surmount the mathematically crippling wall of solving the many-body wave function. According to Cooper, Bardeen "omitted to mention that practically every famous physicist of the 20th century had worked on the problem and failed."[2]

Cooper quickly encountered the daunting and insurmountable equations. On a seventeen-hour trip to New York City, he tirelessly tried his extensive bag of mathematical tricks but got nowhere. He ran out of technical steam and started feeling that the equations were preventing him from seeing the root of the problem. So he decided to step back from the equations and think intuitively about the problem. And then he made a simple and ingenious guess. Part of his mental wizardry was to simplify the problem and avoid unnecessary details, decisions geared toward making the problem more tractable.

As a hint into Cooper's insight, recall that electrons carry a tiny magnetic pole. This pole can also obstruct their motion, say when they are flowing in a current, due to the magnetic forces of surrounding electrons, which cause deflection and electrical resistance. It seemed that, if superconductivity were going to be possible, then the golden rule that electrons necessarily repelled each other had to be broken. Cooper realized that if the electrons could pair up, with their spins oppositely aligned, then the members of each pair would lose their identity as electrons, and their overall spin would vanish, mitigating the local resistance. An emergent phenomenon, the Cooper pair, was born. But the grouping doesn't stop there. When all the electrons pair up, they clump together to collectively behave as one object and move in a ghostly fashion through obstructions in the metal. Cooper likened it to a line of ice-skaters, arm in arm: "If one skater hits a bump, she is supported by all the other skaters moving along with [the line]." In other emergent phenomena in condensed matter, this long-range order is a collective behavior of the individual electrons or atoms.

The formation of Cooper pairs led to a handful of other emergent properties in the superconductor. First, in order for the supercurrent to maintain itself, the superconducting environment would have to expel any magnetic field. This observation is known as the Meissner effect and is predicted by the BCS theory—it is the reason magnets levitate above superconductors. The underlying physics in superconductivity was later found in other systems, such as neutron stars. The extremely dense environment of neutron stars enables neutrons to Cooper pair and exhibit the collective behavior of a superconducting fluid, called a superfluid. Another Nobel Prize was awarded to Yoichiro Nambu, who applied the BCS theory to understand the emergence of particles called the pion, which was found to be a Cooper pair of quarks. BCS theory also inspired some of the architects of the standard model of particle physics to think about how mass could emerge from a similar type of symmetry breaking, and we will discuss this in an upcoming chapter.

In a seminal essay entitled "More Is Different," Nobel laureate Philip Anderson puts emergence at center stage over reductionism in physics: "The ability to reduce everything to simple fundamental laws does not imply the ability to start from those laws and reconstruct the universe. . . . At each stage of [emergence] entirely new laws, concepts, and generalizations are necessary." Anderson goes on to identify the organizing principle that is behind most emergence in condensed-matter systems, that is, symmetry breaking. When we see symmetries, we often see an underlying pattern of phenomena. For example, in relativity, the space-time symmetry inherent in the laws of motion functions to give relative lengths and time for different moving observers. Emmy Noether proved that symmetries are linked with conservation laws. And symmetry breaking signals new properties that are hidden from the symmetric realm.

To see this, consider a piece of metal, like iron. Like a checkerboard is a repeating arrangement of black and red squares, all metals are repeating periodic arrangements of atoms with electrons waiting to easily flow from one site to the other. At each atomic site the electrons carry a tiny magnetic pole due to their quantum spin. Recall that the quantum spin can either be up or down, reflecting its quantum nature. At high temperatures the billions of the electron spins point in a random direction. If we add all the spins, because of the random orientations, the total spin of the system cancels out to zero. Since the total spin and magnetization is zero there is no preferred direction of all the spins and the system has a symmetry that is invariant under rotations. This means that the iron has a symmetry that is analogous to the symmetry of a sphere. However, as we lower the temperature to a critical value, the spins all on average spontaneously pick a direction, attaining a net magnetic field. This is like balancing a pencil on its tip. As a result, the spherical symmetry is broken by the emergence of magnetism. Not only is magnetism emergent, but the preferred direction that breaks the symmetry creates rigid directions that can support the transmission of magnetic waves. In all the examples in condensed-matter physics, the constituents, such as electrons or other fermions, exhibit cooperative

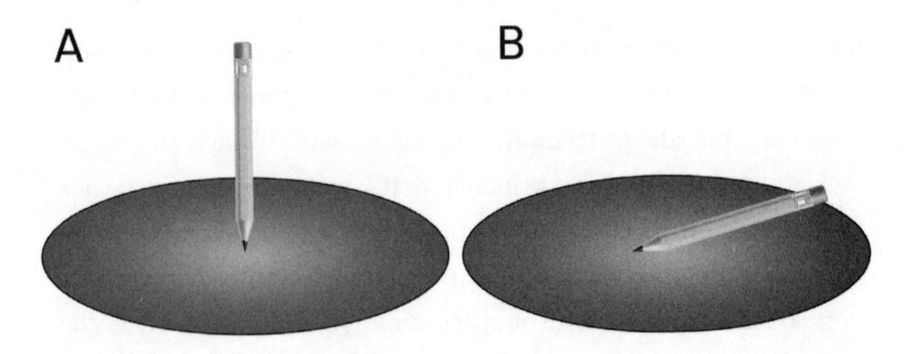

FIGURE 10: An example of spontaneous symmetry breaking. In image A the forces acting on the pencil have a rotational symmetry. However, this system is unstable because the gravitational force in the z-direction will break the symmetry. In image B the pencil picks a direction, which results in breaking the original rotational symmetry. Any direction could have been selected, and this randomness is a sign of the symmetry being spontaneously broken.

behavior similar to the swarm of ants or bees that collectively perform a task that an individual member cannot accomplish alone.

In hindsight it was also discovered that superconductivity emerged from the breaking of another type of symmetry, not related to space-time but closer to those found in the fundamental interactions, such as the strong and weak nuclear forces. The principle of emergence and symmetry breaking turns out to be at the heart of important matters in the fundamental forces. As we will explore in the next chapter, the nuclear interactions, such as the weak and strong forces, are governed by symmetries associated with the charges of elementary fields and particles. And the breaking of these symmetries also has essential properties, such as the origin of mass and the emergence of matter over antimatter in our universe. Even stranger is the idea that space and time are also emergent properties. By analogy, these are atoms of space and time whose collective behavior can give rise to the malleable space-time fabric that Einstein discovered.

There has always been a deep interplay between physics and other scientific disciplines like biology, chemistry, and the social sciences, where even more mysterious forms of emergence occur. Does the organizing principle of symmetry breaking seen in physical systems apply to understanding emergence in other domains? Are there other organizing principles that go beyond symmetry breaking? And maybe even beyond physics?

6

IF BASQUIAT WERE A PHYSICIST

The wisdom of John Bardeen, a two-time Nobel laureate, to seek and integrate the outsiders' perspective of Leon Cooper not only enabled cracking the code of superconductivity but opened the floodgate for breakthroughs in other branches of the physical sciences and technology. If we want to catalyze more solutions to the current mysteries we face, we could try to replicate these examples of scientific inclusivity that Bardeen and others exemplified. These days there is a big push for and rhetoric surrounding diversity and inclusivity in science. But science has fallen short in benefiting from effectively embracing outsiders, or even wanting to do so. Science is carried out by individuals and groups of individuals, and the principle of emergence also acts on scientific societies sustaining forces that act to prevent the full benefits from the contributions and presence of outsiders. In this chapter, we explore through the lens of the science of groups of people—the science of sociology—to see how the scientific community and individuals can inspire and enhance more innovations and innovators. To gain insight into these issues, I present two stories.

SCENE 1

Eleven years ago, Jim Gates, a theoretical physicist with a Frederick Douglass–style 'fro, was the chair of Howard University's physics department. I considered going to Howard as an undergraduate; it is a prominent, historically Black university with famous alums like Thurgood Marshall, but I opted instead for a private Quaker college. During that time, Howard University made a bold move by poaching a handful of Black physics professors from prominent, predominantly white institutions, and this inspired Gates to move from the University of Maryland. During his time at Howard University, Gates and his colleague Hitoshi Nishino were researching a special symmetry called supersymmetry. Supersymmetry is a theory that attempts to connect the fermions, such as electrons, of our world to the force carriers, such as photons. Although in our experience we see both these fermions and bosons, as the force carriers are called, Gates and Nishino were thinking about a class of particles that we don't typically experience, known as anyons. Anyons only exist in two-dimensional systems, and they emerge from matter particles, but they enable the violation of a fundamental law that matter can only have integer or half-integer spins. Gates and Nishino's work was aimed at developing a theory that linked the supersymmetric relationship between fermions and bosons to anyons. Recall that all observed matter has either integer or half-integer quantum spin.[1] For example, the electron, a main building block for molecules, has half-integer spin. Gates and Nishino wrote a beautiful set of equations that made these anyons supersymmetric and discovered a surprising feature that the equations were conformally invariant—we won't go into the details here, but it's a symmetry that related the microcosm to the macrocosm.[2] The result was published in a respectable physics journal and enjoyed a modest number of citations.

In the subsequent eleven years, a theory developed from string theory called M-theory, which attempts to unify all forces, underwent a conceptual revolution—it encoded a property called holography. A holographic theory is one where gravity is encoded in another theory, operating in one fewer dimension, without gravity. For example, gravitational physics in our three spatial dimensions could be holographically encoded in a two-dimensional theory with no gravity. This work was famously christened the ABJM theory, named after the authors, and was, and is, considered one of the most important results in theoretical physics in decades. The four authors much later discovered that the exact supersymmetric and conformal invariant equations that Gates and Nishino derived had a holographic description in M-theory. Many physicists, including me, were not, and still are not, aware of the original equations of Gates and Nishino. To be fair, the authors of ABJM actually cited the Gates-Nishino work, yet it still did not rise to the acclaim that it may have deserved. After all, the equations that ABJM used were the same that Gates and Nishino had derived eleven years earlier. Why did the community not call this theory Gates-Nishino-ABJM? Why did the community not notice the importance of the original Gates-Nishino work?

SCENE 2

After a few months into my second postdoc, I stopped going to my office to work. The dozen or so postdocs in the theory group were very interactive. However, time after time, I found my attempts to interact with my peers were not reciprocated, and were even ignored. One day a good friend, Brian Keating, who was a postdoc at Caltech, was visiting our group. Brian, who is white, pulled me aside and said, "I know what's going on. I know why you're not coming to your office. I overheard a conversation with some other postdocs, and they said that they want to punish you." So, what did I do to them that would warrant punishment by shunning me? My friend volunteered the reason: "They feel that they had to work so hard to get to the top and you got in easily, through affirmative action." I must admit harboring both disdain for and envy of my postdoctoral colleagues. Most of them grew up with privilege that I did not have and a sense of entitlement that the enterprise of science belonged to people like them. I also presumed that their relationship with physics was different from mine.

For me physics was literally a tool for survival. Reading physics books and solving problems kept me away from the streets of the Bronx. Physics paid my way into college and graduate school, and unlocked the shackles of a likely life of poverty in the Bronx. There were times when I would stay up all night playing with physics problems and equations. Yet, I remain thankful to Brian, because, after that day, I made sure that most of my publications and research over the next three years as a postdoc were independently administered and authored. I did not want to be further penalized by colleagues and peers since I could not change their perception of my not deserving admittance to their elite club. By doing independent work, I felt that it would address the perception of whatever shortcomings may inhibit my future employment. While it was useful to learn how to complete independent work, my strategy still did not erase the perception and treatment that I would receive from colleagues throughout my career.

Because I lacked a feeling of belonging in the group, and so that I could continue to be productive, I moved most of my calculational operations to a café across from Stanford's computer science department and I worked by myself. In hindsight, this isolation was a blessing in disguise; it forced me to develop my outside-the-box thinking. At the time, the discovery of dark energy—a concept closely related to a parameter in the general theory of relativity known as the cosmological constant—prompted the entire group, including me, to rethink the very foundations of theoretical physics. And because I had just spent two productive years in London at Imperial College, where I had developed an improvisational and visual style of approaching problems coupled with Jungian dream analysis that Chris Isham had trained me in, I was very prepared for some outside-the-box thinking about the issue of dark energy. Once I had developed a clear idea of what I wanted to pursue with those techniques, then I would unleash an arsenal of traditionally mathematical devices that I mastered to develop a model. I naturally kept this atypical research strategy to myself in fear of further stigma.

One morning after forcing myself to go to campus, I was struck by an insight that had come in one of those dreams I had told Isham about a year previously. The dream inspired me to focus on the particular properties of space-time called discrete transformations. You can see an example of a discrete transformation when you look in a mirror, which makes your left hand look like it is the right hand of your reflection. I had a sense that I was onto something important, so at my café office, and with a celebratory beer, I started doodling inchoate diagrams on a napkin. I had developed this strategy of free play by presenting many competing rough sketches from my mind's eye of the physics I was developing, before committing to any mathematics. At that stage of my investigations, there were no equations, just different sketches that resembled Picasso-like drawings of the physics of accelerating space-times and their discrete transformations—their reflections.

One day not very long after, I noticed the head of our theoretical physics group, Michael Peskin, on a physics chat stroll with the golden-child postdoc of our group, the guy that all the postdocs wanted to be like. As Peskin and the golden boy walked by, I couldn't hold myself back. I said nervously, "Michael, I think I got something interesting about some work on the cosmological constant." Peskin engaged me, and I started blurting out what in hindsight were pretty zany ideas. The golden child smirked and said with a tone of dismissal, "You and your crazy ideas again." In fact, golden boy was one of those whose rejection kept me from hanging out in the theory center. But Peskin saw something and said, "That's interesting, why don't you come to my office next week and tell me more about it."

I did that. When Peskin takes a physicist seriously, he usually throws a question back at them, like a physics koan. After I explained my inchoate idea, Peskin challenged me to do some difficult warm-up calculation that would help my idea take shape. For me, it was important that he took me seriously and challenged me. Over eleven grueling months, with much toil and many calculations, those doodles on a napkin transmuted into a publication in the top journal in physics. We had created a new approach to understanding how matter over antimatter was created in the early universe. The paper opened up new directions in astrophysics and cosmology research; it has been cited over two hundred times.

Michael Peskin is regarded as the "oracle" among theoretical physicists; he is very much an insider. But Peskin is unusual; he possesses a quality that enabled him to appreciate and perceive value in my style of doing physics and in the ideas that I generated that were different from the norm.

Coincidentally, Peskin was an office mate at Harvard in the late 1970s with Jim Gates. At the time, Gates was doing pioneering yet nonconventional work on combining supersymmetry with gravity. Peskin was a young prodigy working in a different field of particle physics. Gates shared with me that Peskin took his work so seriously that he spent ample time learning the daunting mathematics behind supersymmetry and even applied it to his work.

Were the expectations and consequences for doing innovative and transformational research simply lower for Gates, me, and other minorities in science? In April 2020 collaborators from the computer, education, and linguistics departments at Stanford University published results utilizing machine learning that asked the following question: Do women and minority scientists innovate as much as their white male counterparts? After a careful look at the research output and impact of 1.2 million women and minority scientists over twenty years, the findings concluded that women and minorities innovate novel contributions more than the majority group. Innovation involves publishing novel contributions that are used often by other researchers. Novelty enables new connections between ideas that generate knowledge. If those groups innovate more, how do we explain their lack of prominence and promotion in the scientific enterprise? In their paper, entitled *The Diversity Innovation Paradox*, the Stanford researchers explained that "Novel contributions by gender and racial minorities are taken up by other scholars at lower rates. . . . There may be unwarranted reproduction of stratification in academic careers that discounts diversity's role in innovation."

One of the authors of the research article was asked why the innovation from minoritized groups went unnoticed; he responded that "the fresh perspectives that women and nonwhite scholars bring are atypical and can sometimes be hard to grasp, so they get devalued by the majority." A scientist like Peskin is rare: he could see talent when other insiders saw African Americans as deficient and as interlopers.

In what follows, drawing on the two stories I just presented, we will dive deeper into how science and scientists can be better positioned to enable future breakthroughs that would otherwise go unnoticed or disabled. Let us keep in mind that there are many stories like this. We will end with an observation of what the scientific community can learn from the art world to transform itself to do better science.

In a seminal work social theorist Robert Merton observed that better-known scientists get more recognition than a lesser-known scientist for the same achievement—he called it the Matthew effect, which alludes to a biblical saying, "for unto every one that hath shall be given, and he shall have abundance: but from him that hath not shall be taken away even that which he hath" (Matthew 25:29). This is also consistent with the well-known saying, "The rich get richer and the poor get poorer." It can be tempting to explain away Gates and Nishino's predicament with the Matthew effect. According to the Matthew effect, Gates's paper would have likely been more widely cited and recognized had he been the chair at the top physics department, like Princeton, which was the home institution to one of the authors of the ABJM theory. This is partially true, because when he published the work with Nishino, Gates was stationed at a historically Black university that was trying to upgrade its visibility to the physics community. However, the Matthew effect does not fully explain why Gates did not get the credit eleven years later, since he is currently well known in the scientific community: he is the president of the American Physics Society, a member of the National Academy of Sciences, and a winner of the highest award given to a scientist by the U.S government, the National Medal of Science.

Insight into Gates's lack of recognition is tied with my experiences as a young scientist in spaces of high repute, where I was stigmatized and shunned because of social presumptions about my belonging in their cohort. And as we will see in what follows, there is a hidden gem for the advancement of science provided that we are able to clearly perceive the functioning of the blind spots of the scientific hegemony and place value on members who have developed new ways of innovating. To do this, I invite you to delve into the hidden ways psychosociological phenomena impact science.

The various activities of scientific enterprise are carried out by a community of scientists, who form a scientific social structure—a normative order. What are the consequences for science when scientists act within social structures? We know that the institutionalization of science as academic disciplines facilitated its growth and its accomplishments.[3] Is there an underappreciated flip side to this? Can the social order of science generate blind spots and even enable bad faith that prevent a better understanding of the mysteries we ponder? To answer these questions, it is useful to develop some tools of analysis from social theory.

SOCIAL NORMS, CULTURAL NORMS

While some might not have access to the kinds of experience of a Black person in America that would give the intuition of W. E. B. du Bois's notion of "double consciousness," social theory provides some insights and tools to enable us to transcend our collective blind spots to the benefit of scientific progress.[4] My curiosity about my predicament as a scientist and desire to break new scientific ground led me to the works of Émile Durkheim, one of the founders of sociology. Durkheim observed that we all live within social and cultural orders, that our lives are regulated by shared social values and moral obligations that he referred to as the conscience collective,[5] and by shared cultural norms, which constitute meaning, and our commonsensical understanding.[6] Social values validate social norms, which enable expectations that are differentiated according to social position, while cultural norms regulate the nature of meaningful actions. Social norms distinguish between right and wrong, while cultural norms constitute the difference between sense and nonsense, being an insider or an outsider.

When thinking about the normative order of science, we need to distinguish between two sets of norms, cultural and social. The first involves the culture that regulates scientific activity, for example, that theories must be logically coherent and empirically warrantable. These norms differentiate between science and nonscience, as with paleontology and creationism, or cosmology and flat-earth beliefs. While this differentiation is permeable around the edges, the distinction between science and nonscience must be sustained. Social norms are the normative orders that create expectations within specific social circles of scientists; they define the boundary of what is accepted and valued within a specific group of scientists. For example, two competing yet viable scientific theories could vary in validity within different scientific groups on grounds of "taste" or the reputation of the architects of the theory. These judgments are mostly subjective but have consequences that help explain the two stories presented earlier.

DEVIANCE BOTH ACTIVE AND PASSIVE

We have talked about social and cultural norms and how they create the boundary for actions that are considered acceptable. But what about those who violate those expectations? Well, there's a word for such violators: *deviant.* Durkheim posited that the social order is maintained and replicated by the existence of deviance; you cannot have one without the other. And this has big implications for innovation in science. In the first instance, violations of cultural norms make no sense, so we endeavor to make sense of them, often by labeling the violation, or violator, as "crazy." Violations of social norms are "wrong," and to reinforce our sense of what is right, we negatively sanction known violations of social norms. Thus, social and cultural orders are maintained by penalizing deviant behavior. When social and cultural deviance are punished, the punishment constitutes or reinforces the boundary between what is allowed and what is disallowed.[7]

We need to be careful about the use of the word *deviance,* because it actually has two meanings. One meaning, the usual one, connotes a deliberate violation and disruption of the laws. You might think of someone who robs a grocery store. We can call this active deviance. However, the current social order in the scientific community institutionalizes and implicitly promulgates that people of color or women as categories of persons are incapable of doing science as well as white men. You might think of the ways such stigma has followed me through every level of study for reasons that have nothing to do with my actual ability, or the ways that I have described being pushed out of social groups for the crime of existing in the community at all. I am deviant by default. We call it passive deviance. The persistent unwelcoming behavior a scientific community exerts on minorities engenders isolation, which in turn can stymie productivity. The irony here is that passive deviant actors can be a positive asset to the scientific social order; this is, at least in part, because they are more likely to be positive deviants, to violate social and cultural conventions that restrict the bounds of scientific creativity. This is consistent with the research findings in the *Diversity Innovation Paradox* article.

Deviant violations are punished; however, whatever form the punishment might take, this will both discourage future violations and reinforce institutionalized cultural and social expectations. As a consequence, physicists might be motivated to avoid forms of deviance that could result in creative breakthroughs. It is important to distinguish between deviant behavior that is harmful and deviant behavior that results in scientific innovations. Let us explore whether the latter may emerge among minority scientists who carry markers over which we have no control.

Marginal people in disciplines like physics may be in a valuable position to innovate fundamentally because they are likely to expand the plurality of ideas, approaches, and techniques in the discipline.[8] They are less likely than those who "fit in" to feel the pressure to remain within the constraints of their discipline. In my case, though I had the same technical training as my postdoc peers, my social isolation from the group enabled me to both not replicate conceptual blind spots and to embrace ideas on the fringes of established knowledge. But how could science emancipate itself from this fate of suppressing contributions from outsiders?

As a youngster in the Bronx, I lived two blocks away from the last stop on the 2 train, which served as a depot for trains to be serviced. During the sleeping hours, regardless of the weather or unseen dangers, graffiti artists would gather at the depot to work on their masterpieces; my favorite was the larger-than-life spray-paint portrait of the Marvel comic archenemy of the Fantastic Four, Dr. Doom, which covered an entire train car. The 2-train depot served as the local art gallery for schoolchildren as we waited for the daily yellow school bus. I and many of my friends were enchanted by comics and drew our own characters; we looked up to the enigmatic graffiti artists as intrepid heroes, modern-day Peter Pans who broke the rules to do their art.

Painting graffiti on the subway came with the penalty of a fine or criminal arrest. Graffiti expression walks a fine line between art and vandalism. If a graffiti artist does not have permission, then the art is deemed illegal and is considered vandalism. Andy Warhol, an insider in the New York art scene, gave graffiti and street artists permission and validation to have their graffiti integrated into the art establishment. In the 1990s Brazilian graffiti artists were harassed and sometimes shot at by the police. Ironically, today Brazilian graffiti has led to founding art schools in low-income neighborhoods and to a collaboration with police to paint murals in devastated areas. These days, you can find graffiti art in the most prestigious art museums and galleries on the planet; the "best" work sells for astronomical amounts.

There are some valuable lessons that science can learn from graffiti. The graffiti artists benefited the art establishment when they embraced their outsider status and continued making graffiti independent from the hallowed art galleries. For example, post-graffiti artists like Banksy, Samo (a graffiti duo of which Basquiat was a part), and Keith Haring established themselves, and continued to work on the outskirts, while gaining mainstream acclaim. Renowned graffiti writer Eric Felisbret says it well:

> From the perspective of a graffiti artist, the debate about whether graffiti is art or crime is pointless because, ideally, it is both. In the graffiti community a writer cannot achieve status solely based on artistic ability. The writer must also be willing to work outside the law and assume great risk. The movement—which I have been documenting in New York for over 30 years—was founded on this principle and it defines its essence.[9]

Outsiders who craft nonconventional ideas and develop new techniques, similar to graffiti, can be seen as vandals, and that "vandalism" may be penalized. Innovating outside the mainstream is hugely risky. However, the realization that some forms of deviance result in positive accomplishments was a game changer for me. The sense of alienation I felt in science, with all its rejection and stigma, also comes with the advantages of being an outsider. There is value in having an outsider's perspective and an opportunity for innovation from being in my natural state when I am taking intellectual risks. In 2018 I wrote in an essay:

> I've come to realize that when you fit in, you might have to worry about maintaining your place in the proverbial club. There are penalties for going elsewhere or doing things your own way, as nonconformity can feel threatening to the others in your circle.

So, I eventually became comfortable being the outsider. And since I was never an insider, I didn't have to worry that colleagues might laugh at me for an unlikely approach. Many times, that unorthodox approach actually led to new understandings.

Both the art world and the physics world have deviant actors. The difference is that the art world has embraced graffiti, while the scientific community has yet to embrace those who take risks. It is clear that embracing graffiti was good for the graffiti artists, but I'd argue that it was equally good for the conventional art world. Imagine contemporary art without Basquiat's beautiful and unsettling figures, or Banksy's consistent challenges to authority in both art and politics. Imagine city streets in which all the street art and the brickside murals are painted over. I don't like to imagine a world without the richness and the beauty that those contributions brought.

I like to imagine Basquiat as a physicist. I think of him strolling down his university's hallways, maybe stopping briefly to chat with his colleagues. I think that the blackboards in his offices would be covered with drawings as well as equations. I think that students would come to peek at the art while he worked. If Basquiat were a physicist, his work would be as unconventional as his office. I expect that he would break the rules, and as graffiti artists do, he would take pleasure in doing so. And those students who came to watch him working, I think that they'd learn how to break the rules too. If Basquiat were a physicist, he would be able to recognize immediately the value of contributions that others in the field might see as simple heresy. And I'd like to think that a physics community that welcomed Basquiat into its fold would want to take the risks associated with giving such contributions a fair shot, even if the rest of the community might not immediately see the value in Basquiat's work. I think such a community would be richer for his presence, and, for his presence, able to grow richer over time.

A naive conversation about diversity in sciences, often filled with gesticulations of identity politics, sustains the smoke cloud that obscures the real issue, the true value of not completely belonging, of not always being comfortable around others, a discomfort that difference brings. Assuming that outsiders attain the competencies of the field, their outside perspective and nonconformity can be exactly what is needed to facilitate major breakthroughs. In fact, many significant innovations in science came from someone who was an outsider in a given field, someone who applied a new technique or perspective from another field. Perhaps this was because they were valued within both disciplines. Perhaps it is time to value and elevate minorities, thus enabling them to make major contributions, not in spite of their outsider's perspective, but because of it.

PART II

COSMIC IMPROVISATIONS

Cosmic Improvisations' narrative and structure will proceed like a real-time jazz improvisation, wherein readers and I will solo on some of the most pressing issues and controversies in fundamental physics and cosmology. To this end, I have structured the rest of the book like a jazz album. In the jazz tradition, a tune often begins with the "head," the main melodic theme. The head is connected to the harmonic and rhythmic structure—the form—of the piece. A soloist improvises over this form, using the head as a context in which to search for and discover new melodic ideas. Saxophonist John Coltrane's improvisations exemplify this process: he integrated and creolized musical forms that were thought to be incompatible—as in, for instance, his masterpiece, *A Love Supreme*. He broke with tradition as he transcended it, incorporating in it the new musical styles he created. Using my three principles in Part I of the book, we'll explore how new ideas in physics are created, and try to answer some of the biggest puzzles in cosmology—including the nature of the big bang, the cosmic origin of life, and the role of consciousness in the universe, as well as discussing the possible quantum nature of gravity, and the role of dark matter and dark energy in cosmic evolution.

And we'll do this drawing on my improvisations and work, as well as those I've undertaken with my collaborators. But we'll also look at the work of other cosmologists and scientists. While improvisational logic is key to guiding us to new landscapes in physics, it's not enough; as every jazz musician knows, we must also learn from the improvisations of others.

7

WHAT BANGED?

The universe is expanding. Why should we care? After all, it seems like the expansion has no effect on us, right? We are bound by gravity to our solar system, and if you look out at the night sky, you see stars, no different from our own sun. Like grains of sand on a beach, our sun is as typical as the other stars in our galaxy, filled with hundreds of billions of suns. But the expansion of the universe matters to those stars, which means it matters to us. Stars are necessary not only for sustaining life as we know it with their light and heat, but for functioning as the manufacturer of the material of planets and life. As we will soon see, there is a startling connection between the early evolution of the universe and the creation of the substances necessary to form stars, which points to a delicate interplay between gravity and quantum physics. And some of this interplay remains unsolved.

Einstein taught us that space and time are more than the location of an event, but that, like electromagnetism, space-time itself is a dynamical—in its case, gravitational—field. Not only does matter live on space-time, but space-time lives on matter. Our most direct experience of this fact is that the sun warps the gravitational field, like a sitting person's weight warps a cushion, bending space-time and creating orbital contours through which planets can move. In general relativity the space-time field, called the metric tensor, encodes information about the curvature of space-time in the presence of matter and energy. As John Wheeler famously states, "Matter tells spacetime how to bend and spacetime tells matter how to move." But space-time can dance in many different ways depending on the configuration of matter and energy that interact with the gravitational field. A famous example is the spherically symmetric space-time of a black hole that is sourced by a collapsed star. The mass density is so high that space warps such that the emitted light cannot escape the highly curved space-time.

By now it is popular knowledge that the universe is expanding, and this has been experimentally confirmed by looking at receding galaxies and racing away exploding stars called supernovae. When the distribution of matter and energy is the same in every direction and at every point in the universe, the equations of general relativity predict that the universe's space-time will expand. A simple way to intuit this type of physics is to imagine the surface of an expanding balloon with points fixed on the surface. Imagine that all points on the balloon's surface is the environment of a galaxy.

As the balloon expands, the galaxies appear to recede from each other. An observer in a given galaxy is fixed at the same point on the balloon's surface despite its expansion. What's moving is the space (rubber) between the points, or galaxies, on the balloon. If you can imagine that a region on the surface of the balloon is a region of three-dimensional space, then the analogy comes pretty close to illustrating the dynamics of an expanding universe. Where does the analogy break down? The rubber makes up the balloon, but space in our universe seems empty, yet dynamical. It appears that space is continually being created in an expanding universe, and we will discuss this perplexing fact in a later chapter.

But what is the significance of the expanding universe, aside from making our universe extremely large to house billions of galaxies? Even after I wrote my first research paper that involved some new features of the expanding universe, I wondered about that question, but was too embarrassed to mention my puzzlement to others. I soon discovered that a copioneer of the expanding solution of general relativity was a Belgian Catholic priest and theoretical physicist, Georges Lemaître, in 1927, who was driven to reconcile his religious beliefs with his love of and conviction for the veracity of general relativity.[1] From this solution Lemaître predicted that the expansion would make galaxies recede from each other at a speed that is proportional to the distance between them. Two years later Edwin Hubble confirmed this prediction by showing with telescopic data that the light emitted from galaxies was redshifted as predicted by Lemaître's expanding solution—indicating that galaxies are moving away from each other and us.

Lemaître seemed to have been able to reconcile his religious conviction with the materialist explanation from relativity theory by positing that the universe emerged from an initial point of "creation" where all the matter in the universe was concentrated. He called this point the "primeval atom" or the "cosmic egg, exploding from the moment of creation." This view of the creation of the universe is also found in many proto-Indo-European and African cultures. This later became known as the big bang, coined by English astronomer Fred Hoyle, who was a proponent of a competing theory coined the steady-state universe. Lemaître says it quite poetically: "We can compare space-time to an open conic cup. . . . The bottom of the cup is the origin of atomic disintegration; it is the first instant at the bottom of space-time, the now which has no yesterday because, yesterday, there has no space."

According to Lemaître's solution, distances increase as time in the universe elapses. If we run the cosmic clock backward then there will be a time in the past when all distances tend to zero, leading to the nonexistence of space. The geometry of this expanding universe can be recast in a mathematic form that the universe's beginning looks like a point in a four-dimensional surface that spreads into a cone in the future. The priest intuited that the entire content of the universe was contained at this point of origin, the "bottom of the cup" that he called the "primeval atom" that burst into an expanding cone. However, he did not provide a mechanism for this disintegration of the primeval atom, nor what conditions led to this initial state of the universe. And there is another serious problem associated with this state of infinite density. As the universe shrinks, the curvature tends to infinity. When physical quantities of a theory go to infinity, we call it a singularity and it usually means that the theory itself breaks down and can't be trusted. But there are some subtleties about cosmic singularities that deserve more attention and background, which we will discuss in a later chapter. History has shown that singularities signal new physics that may resolve the singularity. But there is another clue that may give us insight about the very birth of our universe's space-time and what may have sourced it.

When the expanding universe was smaller and denser, it was much hotter. Starting from the big bang as the universe expanded, roughly one hundred thousand years later the temperature cooled such that its energy was enough for electrons and protons to bind, forming hydrogen. During that moment the last photons would be liberated into space, leaving a thermal afterglow at the energy associated with the binding energy of hydrogen. This is the last fossil light of the early universe that would persist to propagate throughout the universe with a temperature of three thousand degrees Kelvin. But as the universe expanded by a factor of one thousand since the CMB epoch, we should expect to find this light radiation in every direction at this temperature. Physicists were on the hunt to find this afterglow, which was coined the cosmic microwave background radiation (CMB).

FIGURE 11: This is a Penrose space-time diagram of an expanding universe (top) and a collapsing one. The jagged lines represent the big bang (crunch) singularity, and the solid diagonal lines represent the horizon.

Finding it would be a smoking gun confirmation of the expanding universe. The Nobel Prize–winning discovery was made in 1967 by Arno Penzias and Robert Woodrow Wilson, two Bell Lab scientists who first thought that the signal was a contamination from pigeon dung. Down the street at Princeton, physicists such as Robert Dicke and James Peebles were looking for the afterglow radiation from the big bang, but with no success.

When we look in opposite directions in the night sky, we see that the CMB photons each took 13.8 billion years to travel to us. Those antipodal photons would take over twenty-six billion years to reach each other. Because the photons travel at the maximum allowed speed, we can know the largest distance possible the photons could have covered, the horizon. Since those photons all existed at the time the CMB existed, we can ask if they had enough time to speak to each other at some earlier time. The time between the big bang and when the CMB existed is three hundred thousand years; we find that the photons were never able to be in contact with each other. In other words, they are outside each other's causal horizon. How is it that the photons that cannot communicate with each other attain the same temperature? Some unknown physics that seems to break the speed-of-light barrier between the big bang and the CMB, which is about one second in the expansion history, had to take place so that the photons had the same properties. Otherwise, there is a bizarre coincidence that gives these photons the same properties that appear to occur faster than the speed of light. However, this would violate Einstein's principle of special relativity—the speed of light is finite and universal.

But it's not only the CMB photons that seem to have a nonlocal origin. After all, if the universe were completely filled with only photons and electrons, it would be a boring place. How did all the stars, galaxies, and clusters of galaxies come about? In 1989 a more precise measurement of the CMB was made and tiny ripples in photons

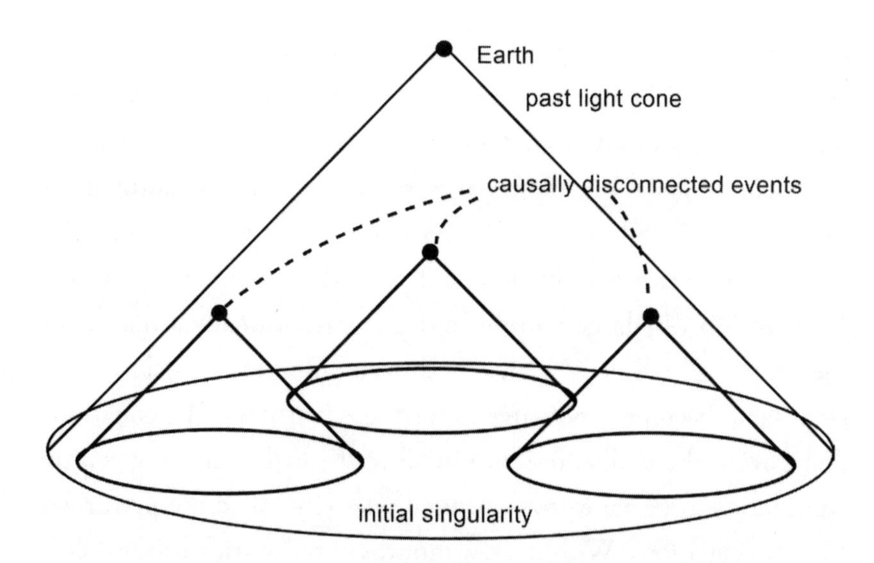

FIGURE 12: A light cone connecting Earth (at top of image) to the big bang (at bottom). Three other light cones specify the histories of three causally unrelated events in the universe's past.

were found. An analysis of the distribution of these ripples showed that they matched the superstructure of galaxies in the universe today—those tiny ripples eventually grew into the stars and galaxies that occupy our current universe. Here is how it happens.

The physics of the CMB plasma is like a seething hot ocean of a fluid that's made up of electrons and photons. This cosmic ocean is dancing with waves that have a special pattern called scale invariance. This scale-invariant pattern has a property that the size of waves of all different frequencies is the same—like zooming in on the plasma and seeing the amplitude of smaller waves look the same as larger ones. A random chaotic fluid does not have this pattern, and some special conditions need to set up an undulating plasma with this type of scale-invariant pattern of waves. But what is the purpose of these scale-independent waves? Let's focus on one such wave.

Plasma waves are nothing but pressure waves and are plentiful in nature. Familiar sound waves in the hollow column of a flute is a pressure wave. When a pressure wave is at its peak the mass density and pressure of the wave is maximized. In a gravitating medium like space, these regions of high density will gravitate more than regions of lower density. While these waves in the CMB are oscillating from regions of high to low pressure, space is expanding and cooling the average temperature of the plasma. Eventually the electrons get captured to form hydrogen atoms and get gravitationally attracted to the waves of high density. The hydrogen atoms start to cluster together and become candidates to form the first stars. The equations that govern the oscillations and infall of the hydrogen are given by considering a special form of general relativity called the perturbed Einstein equations. Within a few months of me writing this, my colleague Jim Peebles won the Nobel Prize in physics for solving these equations and illuminating the correct physics that led to the formation of the first structures in the universe from this nearly scale-invariant spectrum of the CMB's waves.

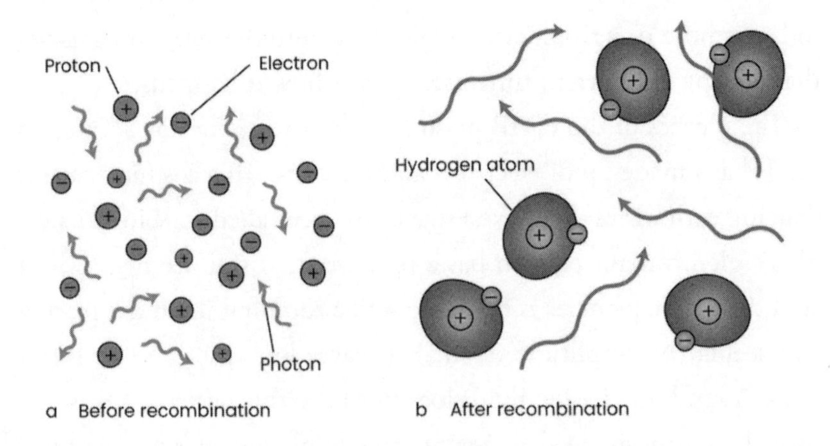

FIGURE 13: To the left, during the recombination, epoch electrons and photons exist in thermal equilibrium. They have the "fluidlike" properties of a hot plasma. To the right, after recombination, the universe cools and electrons become bound to the protons, forming hydrogen, liberating the CMB photons.

The details of how this happens make a more complicated, but straightforward story. It turns out for Peebles's equations to correctly describe large-scale structure of the universe, the CMB also will have to have extra gravitational pull from overdensities of a new component of matter, namely dark matter.

Unlike the plasma waves, dark matter is required to have zero pressure. If this dark matter overdensity has no pressure, then its oscillation will come to a halt almost immediately. This is because pressure waves, like springs, resist compression and push back against compression with more force the more they are compressed. These two opposed forces of compression and resistance create periodic oscillations. On the other hand, when the dark matter settles down into various regions in the sky, the nearby hydrogen will experience even more attraction toward the dark matter overdensity, enhancing its tendency to form structures, such as stars and protogalaxies.

There's a major problem with this picture, however. The spacetime structure of the expanding big bang universe does not allow enough time to generate the pattern of scale-invariant waves. So, something special had to happen before the CMB epoch that set up all these waves, like a cosmic conductor telling all waves to oscillate at the same time and loudness. For that to be true, the expanding universe and the observed properties of the CMB—it even seems as if our existence—arise from a mysterious nonlocal phenomenon in the early universe, what Einstein called spooky action at a distance. Physics acting beyond the horizon and sensitive to a big bang singularity seems like an inevitable affair to explain our current existence, and this will require us to consider the possibility of nature exploiting some quantum magic.

The magic of dark matter isn't just that it provides the missing gravity necessary to hold the universe we see together. Many of us take for granted that next year will come, because we assume that our solar system's orbit is stable—without dark matter it is not.

8

A DARK CONDUCTOR OF QUANTUM GALAXIES

The energy in the undulations that grew under gravitational collapse some fourteen billion years ago needed an invisible form of matter to efficiently form the billions of galaxies including our own Milky Way.

We can measure the effects of dark matter in different ways. The rotation of stars in galaxies is the most direct. Today we see the striking presence of dark matter in all galaxies, which is inferred from an anomalous rotation.[1] How is this? When we look at gravity at distances comparable to our galaxy, it fails to account for the motion of our sun and solar system around the Milky Way. The speed of a star like our sun around our galaxy is proportional to the amount of mass it contains. From this, we can infer from the sun's velocity that about 85 percent of the mass is missing or invisible/dark. In other words, our sun is moving so fast that if there weren't some hidden form of matter providing the necessary gravitational pull, it would fling off from the Milky Way into oblivion. We have even found that every galaxy is made up mostly of dark matter.

A second way we can measure the effects of dark matter is through a phenomenon called gravitational lensing. Similar to the bent lens of a magnifying glass, a gravitational lens is a region of warped space that causes light that traverses it to get distorted. Imagine that there is a massive blob of dark matter in front of a galaxy. Although we will be able to see a visible galaxy that lurks behind the dark blob, the image of the visible galaxy will appear to be distorted in a definite way according to general relativity. We can infer the mass of the dark blob from the lensed image of the visible galaxy.

So, we now know from observations that dark matter exists in individual rotating galaxies and clusters of galaxies that form a cosmic web of interconnected galaxies spanning cosmic distances. I fooled a neuroscientist who thought that a picture of this structure looked like how the brain is wired with neurons. There is invisible dark stuff that pervades the universe, and aside from being the cosmic glue that keeps stars like our sun in orbit, invisible things have a way of being taken for granted. So why are physicists and astronomers so interested in dark matter? We believe that unveiling dark matter is a puzzle piece that will help us understand the fundamental nature of our physical world. And we suspect that once we crack the dark matter code, we will come to know something unanticipated. After all, did Einstein ever imagine that the quantum nature of the photon would lead to solar cells? Ever since its discovery, we have been cooking up theories to account for dark matter, and despite our efforts, we have not been able to identify the one true model behind its mystery. We are not short of imagination, as there are hundreds of candidates for the identity of dark matter. Let's say that you wanted to construct your dark matter model. There are some necessary criteria that it needs to satisfy.

Our standard big bang cosmology suggests that dark matter may have been born in the very early universe at least fourteen billion years ago along with the visible matter, and based on the observation of the cosmic microwave background radiation, both dark and visible matter were distributed across the universe soon after their creation. After the dark matter is created, it must remain stable throughout cosmic history, meaning that it cannot annihilate or decay so that it can cluster to form the scaffolding for the visible matter to form stars and galaxies. Second, for the dark matter to do its job to aid gravity in sculpting the cosmic structure, it must be cold, or equivalently, the dark particles must be motionless, which endows dark matter with the properties of a pressureless, frictionless fluid. The opposite of this is a gas of particles bouncing around so fast that their high momentum would make them hot, which would not allow structure to form.

As a result, the pressureless dark matter fluid does not exert any preferential flow in any direction, so it would rather stay put—much like an invisible cosmic Silly Putty. The dark matter fluid, during the earliest stages of cosmic structure formation, does undergo localized undulations, creating distortions of attractive gravitational energy in the fluid. A correct understanding of the origin of these undulations of the dark matter is still a matter of research and will be discussed in the chapter on the physics of the very early universe. The energy in overdense regions of dark matter generates a gravitational attraction to enhance its gravitational pull on the nearby visible matter that was around in the early universe, such as electrons and photons. The visible matter will collapse around the center of the location of dark matter density, eventually forming the region for a protogalaxy. The process I just described is more complicated since it is quantitively described by a set of coupled differential equations describing how dark matter sources gravity, and how the visible matter exerts radiation pressure to create oscillations as it collapses.[2] The system quickly becomes nonlinear and requires numerical physics to analyze thoroughly. But the takeaway is that once we specify the dark matter, visible matter, and general

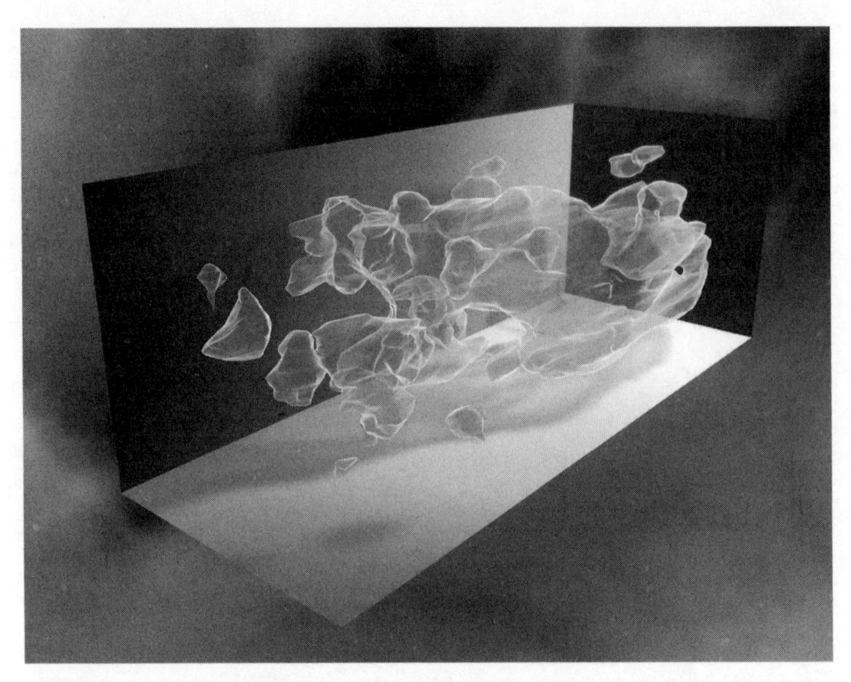

FIGURE 14: An artist's impression of the three-dimensional map of the network of large-scale distribution of dark matter in the universe. Apparent in the map are filaments of dark matter that continue to collapse under gravity to seed galaxies.

relativity, we get the correct predictions for the imprint of this physics in the observed temperature fluctuations in cosmic microwave background radiation.

So how do we make progress given the limited things we know about dark matter and the plethora of theoretical models on the market? Most models fall into two categories. One is that dark matter is simply a new particle that's not found in the pattern of fields in our standard model. This dark particle/field will just interact very weakly with the visible matter. Recall that these patterns follow from symmetries that relate these fields to each other. The other theme for dark matter is that there is no dark matter, but the force of gravity changes in the contexts where we see the effects of dark matter. We call these models modified Newtonian dynamics, or MOND, a hypothesis created by the physicist Mordehai Milgrom.

The basic idea of MOND is that Newtonian gravity is valid on solar system scales. Still, as we go into the regime of galaxies, gravity gets modified to account for the observational fact that all spiral galaxies rotate at a universal acceleration that is proportional to the expansion rate of the universe. Galaxies are gravitationally bound and isolated systems that are decoupled from the expansion rate of the universe. A typical galaxy is on the order of one thousand parsecs, and a cosmic expansion occurs at distances around billions of parsecs. So how could galactic rotation communicate with the universe's expansion rate? The jury is still out, but the scientific community leans toward dark matter being a new particle due to the observation of the bullet cluster—two galaxies that collided with each other—leaving in its wake a separation of dark matter and the visible matter of the two galaxies. This observation is tough (but not proven to be impossible) to be explained with MOND.[3]

During most of my college and graduate school years, I was a competitive 800-meter runner. I trained religiously for three hours five days a week to meet my goal of finishing the half-mile in one minute and 51 seconds. Over the years, I would coincidentally see the number 151 appear when I looked at the time, the number on a billboard, and many random places.

The number 151 seemed to appear at a frequency greater than any other number. I took this coincidence as an indication that one day I was bound to run that time, but twenty years have gone by and I still haven't. These days my knees would not even get me through four hundred meters in two minutes. Had I not given so much significance to the coincidence of the number 151, I would have settled my goals on a more realistic 154 for the 800. But aside from sports and gambling, the seduction of coincidences also creeps up in physics.

One of the themes in attempting to solve a problem in theoretical physics appears in coincidences. In the dark matter world, a few coincidences are floating around. When we try to make progress using a coincidence, it's crucial to be mindful of our tastes and prejudices. One of the most popular coincidences led to the possibility that dark matter could be detected directly from its interaction with matter subject to a weak nuclear interaction. The early universe is a hot thermal soup of particle species. If dark particles exist and interact, then they will also exist as a thermal soup as well because the density increases in the past universe. Most particles, including dark particles, come with their associated antiparticle that can annihilate into lighter and more stable particles like electrons and photons. Still, this process is suppressed as the universe cools and expands, since it is less likely for particles to find their antiparticles as they run away along with the expanding universe. There is a beautiful relationship between how the decay of a heavy particle into a light one will depend on the mass and energy, and this applies the same way for both dark and visible matter.

What is the lightest mass and interaction strength necessary so that the decaying soup of heavy, dark particles can give the correct density of dark particles in the universe today? It turns out that the mass of the dark particle is related to that of the weak interaction of visible matter. This is a numerical coincidence and is famously called the WIMP (weakly interacting massive particle) miracle.

There is another good reason astrophysicists took the WIMP miracle seriously. One of the great successes of the standard model of particle physics is the Higgs mechanism, which endows mass to all the particles known to us. The Higgs field accomplishes this feat by breaking the symmetry associated with the weak interaction at the energy scale commensurate with the Higgs particle. A big reason for building the Large Hadron Collider was to search for new anticipated particles at the mass scale of the Higgs mass, which would give insight into the nature of electroweak symmetry breaking and the physics beyond the standard model.

WIMPs have created excitement because of their prospect for being directly detectable in particle colliders and other creative forms of direct detection here on Earth. For example, every once in a while, a WIMP can travel to Earth from a galactic halo and interact weakly with an atomic nucleus, which could give a detectable signal according to the specific type of WIMP particle and its interaction with the nucleus. My colleague Rick Gaitskell at Brown has been one of the pioneers of the quest to detect WIMPs and other dark particles directly. It's been about forty years, but so far, and with ingenious

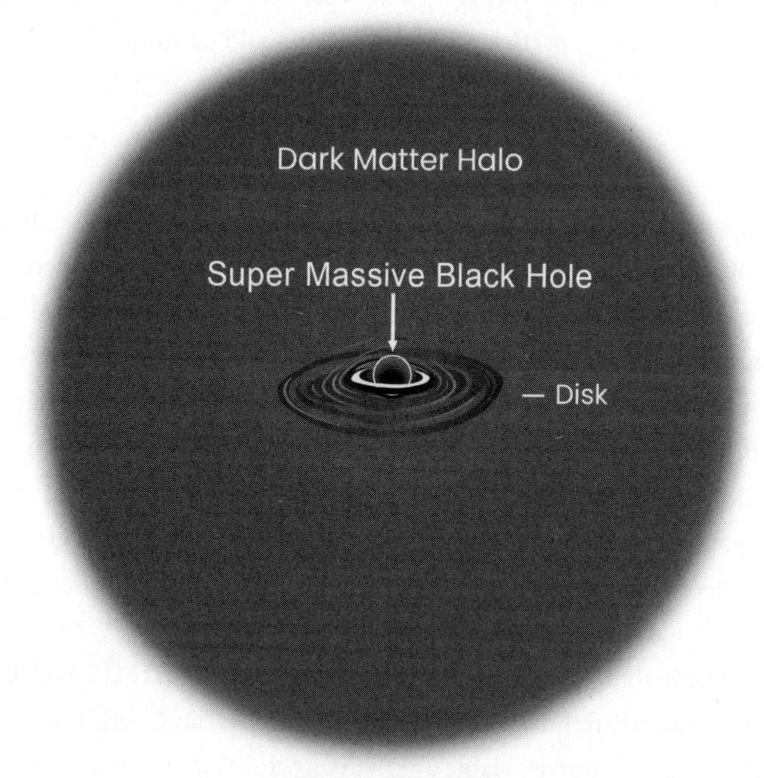

FIGURE 15: A schematic representation of the dark matter halo surrounding the Milky Way, with a supermassive black hole at the center.

and laser-sharp accuracy in detecting the faintest interaction with WIMPs, there has been no direct detection. These direct-detection experiments have allowed us to highly constrain and rule out a large class of WIMP models. Still, a number of models, predicting a variety of possible particles to fill the role of WIMPs, survive, and the hunt is still on to see which of those particles might actually exist. There are some interesting anomalies among the different experiments that suggest that WIMPs may be the culprit. I remember having lunch not too long ago with one of the co-inventors of the WIMP miracle model. She lamented to another dark matter theorist, "It's been over thirty years; where are the damn WIMPs?" So, let's assume that the dark matter is not a WIMP. Then what are we left with? For one thing, if it is not a particle, then what else could be in the sky?

One uneventful fall day after teaching my general relativity course, I noticed an email from my colleague David Spergel, a pioneering cosmologist, director of the Simons Center for Computational Astrophysics, and now president of the Simons Foundation. He said that he would be passing through Providence and would have a free half-hour to meet. Spergel was one of the first people to notice the WIMP miracle while he was a young graduate student and later with renowned cosmologist Katie Freese figured out an ingenious way to detect WIMPs. We decided to have a quick dinner close to the Amtrak station in downtown Providence.

I knew that Spergel was a tad frustrated that WIMPs had not been detected after decades of relentless searches, and I saw this as an opportunity to pursue viable alternatives with him. Over a course of Malaysian food, we wasted no time, and I asked him: Aside from the WIMP miracle, what was the most pressing observational mystery behind dark matter that needed to be addressed?

He reminded me of another coincidence, an observation that he was among the first to notice. The observed density of dark matter is the same magnitude as the visible matter density—and a fundamental explanation was lacking.[4] This coincidence is even more striking because visible and dark matter don't seem to interact with each other.

After going back and forth, we realized that the link between dark and visible matter could be addressed by their common origin in the early universe. The realization was fueled by the need to combine the quantum principle with the gravitational side of dark matter. As we will soon see, not only does dark matter gravitate but gravity plays a key role in the origin of dark matter, and this would have problems without quantum mechanics. A part of this issue was pursued and addressed by particle physicist David Kaplan in the 1980s.[5]

Kaplan is known to be one of the most creative particle physicists in his generation, and his idea called asymmetric dark matter was the first to connect creation of matter over antimatter to the creation of dark matter, attempting to explain the coincidence between their densities in the universe. But, how could it be that dark and visible matter can have a joint creation without ever "knowing" about each other? Plus, it is not enough to provide a candidate for dark matter that solely explains its properties in the current universe. It is useful to understand baryogenesis itself to answer the potential link behind dark and visible matter. Even if we were to understand the origin of the dark matter, David Spergel and I understood that our model was calling for something that was not a WIMP. It is first useful to understand how dark matter can be cocreated with the visible matter, otherwise known as baryogenesis—the origin of baryonic matter, such as protons and neutrons.

As it stands, there is no consensus among cosmologists about the creation of matter over antimatter in the universe. But there are some conditions that most of us agree on, which have to do with the known symmetries of the vacuum state of the quantum fields that give rise to all the matter and antimatter. The critical insight into baryogenesis is that special symmetries that related matter to antimatter had to be desecrated by new physics that operated in the very early universe. And as we will see, these symmetries are not cherished in our universe. In 1967 the Russian physicist Andrei Sakharov provided the necessary ingredients to accomplish baryogenesis—these were famously known as the Sakharov conditions. These conditions fit into the principle of emergence that I discussed earlier, where we see that the violation of invariance leads to a variety of complex, emergent, and physical properties. Baryogenesis transpires to be a most crucial cosmic example of the emergence of matter in spacetime, but exactly how the universe does it remains a mystery.

To create matter over antimatter in the universe, dark or visible, Sakharov imagined that three symmetries would have to be disrupted. The vacuum of the quantum fields in the standard model has a symmetry that preserves the number of particles. So, if the universe starts empty, then it will remain barren unless there is a process that violates the conservation—this is called baryon number violation, or B. But even if we have a process that violates B, nothing stops that same process from producing antibaryons. This is seen in the following example: An electron and positron both democratically interact with a photon. If enough energy in the photon is imparted into the electron's field, then its vibrational energy will enable the creation of an electron from the vacuum. But, even if you were a demigod trying to create a certain number of electrons in the universe this way, you would necessarily create just as many positrons as well. But if you want to create a universe like ours, there needs to be some way around this. This is Sakharov's first condition: the baryon number cannot be conserved.

There is another symmetry between the electron field and the photon that also has to be violated. Called CP, where C and P stand for charge and parity inversion, this symmetry is similar to a mirror reflection. When you look in the mirror, there is a spatial reflection that inverts left and right—your left hand looks like your right hand under a mirror reflection. We can imagine CP inversion as a kind of mirror that reflects the charge and orientation of the electron. A spinning particle under a CP inversion sees itself looking like its antiparticle, the positron with its spin reversed. So, if we wanted to create matter over antimatter this CP symmetry would have to be violated. This is Sakharov's second condition.

The third condition is that both charge asymmetry and particle creation would have to occur out of equilibrium, meaning that creating a particle would have to happen such that the particle would not encounter its antiparticle as they would demolish each other. But when in the history of the universe did these conditions occur

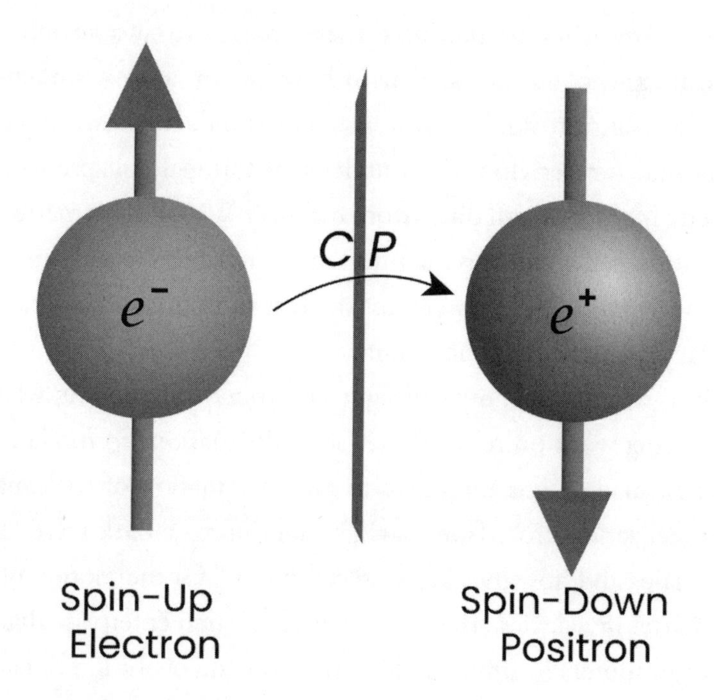

FIGURE 16: Representation of CP symmetry violation in a pair of electrons.

for creating matter over antimatter? Attempting to understand the origin of both visible and dark matter points us to an earlier epoch in the early universe called cosmic inflation that had the correct physical ingredients to realize Sakharov's conditions and a universe that is void of matter for all intents and purposes. We will get to how inflation works its magic creating "something" from "nothing," but let us return to our discussion about dark matter genesis.

David Spergel and I teamed up with my postdoc and fellow cosmologist Evan McDonough to develop a new theory of dark matter that realizes the coincidence between dark and visible matter baryogenesis while being agnostic about the underlying field theory. We let a quantum consistency condition, known as anomalies, dictate what the nature of the dark matter would be and discovered two pleasant surprises.[6] First, we discovered that the dark matter originated from a mirror world of the quarks in our visible world—dark quarks. After they are produced these quarks are close to zero temperature and condense, similar to how helium atoms condense to form a quantum fluid known as a superfluid. The dark quantum superfluid further clumps up under gravitational collapse to make galactic blobs that will differ from ordinary WIMP dark matter. We were surprised about this quantum mechanical result that our calculations revealed, but we were not the first to resort to dark matter as a galactic quantum phenomenon.

Two decades ago, my colleague and friend Wayne Hu, who did pioneering work on relating the CMB fluctuations to fundamental physics, made a quantum leap in our imagination of dark matter. This realization arose from paying attention to a dark secret about dark halos and how they act as a "conductor" for the motion of visible matter in galaxies. There is an entire research enterprise that uses supercomputers to simulate the formation and evolution of galaxies spanning billions of years. These so-called large N-body simulations computationally study the cosmic history of dark matter particles that lead to their formation into dark halos. Time and time again with unprecedented accuracy, large N-body simulations reveal that the distribution of the dark matter in these halos has a cuspyness in its concentration around the center of the galaxy, in disagreement with observations of real halos, whose dark matter distribution is smooth. Hu took this discrepancy seriously and asked what it would take to prevent the dark matter from clumping in the center of galaxies. And his answer was both simple and struck a conceptual resonance with how Newton connected the falling of the apple to the motion of the moon. Hu saw that the same wavelike property that stabilized the electrons' orbit in the atom can prevent dark matter from forming cusps, if the dark matter had a quantum wavelength on the scale of a galaxy!

Recall that the de Broglie hypothesis says that a massive quantum particle should have a wavelength inversely proportional to its mass. This means an extremely light particle will have a long quantum wavelength. The reason we don't see the wavelike properties of particles like electrons and protons is because their masses are large enough that their de Broglie wavelengths are microscopic. What Hu argued is that the dark matter can have a quantum wavelength on the scale of a galactic distance provided that it is light enough. And like a galactic electron the light dark matter will not collapse to the center of the galaxy.

Unlike the plethora of possible WIMP particles, so far there are only two potential candidates for superfluid dark matter, a particle that we'll discuss later called the axion and the dark quark condensate. We are still trying to find new, clever ways of unveiling the true identity of dark matter, assuming that it is a form of matter that doesn't interact with matter that we see. In the case of superfluid dark matter, we can exploit its quantum properties as a way of distinguishing its properties revealed in the sky. And one unique property of superfluids is that they produce quantum excitations called vortices, which have a mathematical pattern similar to how clouds spiral to make the eye of a storm. If galaxies have vortices in them then they would affect the gravitational lens effect and lead to an anomalous observable signature. Currently, research groups, including my own, are using machine learning to hunt for substructure such as vortices within dark halos as a smoking gun signature of the quantum nature of dark matter.

9

COSMIC VIRTUAL REALITY

The ancient idea that the universe was created from a state of emptiness is captured poetically in the book of Genesis: "And the world was without form and void, and darkness was on the face of the deep. And God said, Let there be light, and there was light."

Similar sentiments were expressed earlier in ancient Babylonian, Sumerian, West African, and Maori creation philosophies. Our precision physics of quantum field theory and general relativity combine to give a picture of the early universe that resonates with these creation stories—the idea that everything emerged from the void. The notion of emptiness, of blackness, according to modern physics, is not how we imagine it to be based on our direct experience of empty space. In every region of empty space, quantum fields are seething with activity so rapid that our ordinary-day perceptions are blind to it. But empty space is much more interesting that our psychological and cultural projections make it to be. The modern picture of the early universe is that the ignition of the big bang emerged from a previous primordial state in which the world was devoid of matter in a vacuum state. My goal here is not to turn our discussion into one of a perennial matter but to motivate the underlying physics that presently attempts to explain the creation of the physical universe from an empty state during the big bang epoch, and discuss some current mysteries cosmologists face.

To date, the most accepted and experimentally compelling paradigm of the early universe is cosmic inflation. The critical message of inflation is that every bit of structure in the universe, including planets and living things, emerged from quantum fluctuations from a vacuum state that contains the inflaton field's potential energy—the inflaton is called that because it drives inflation—and nothing else. Nevertheless, what physicists call virtual particles can emerge spontaneously from the vacuum, and these are important to explaining how both matter and large-scale cosmic structure emerged from nothing. And although there are alternatives to cosmic inflation, such as the big bounce and cyclic cosmologies, which attempt to alleviate the problems that plague inflation (we will discuss some of these alternatives in the next chapter), none of them are free from the influence of virtual particles. No matter which way we slice it or which model we prefer, the idea that all matter and structure in the universe came from virtual quantum processes is inescapable, counterintuitive, and requires a more in-depth examination. After all, nature has to figure out how to make virtual particles real particles. And indeed cosmic inflation has a clever trick up its sleeve to make this happen.

In quantum field theory, virtual processes are ubiquitous and have been measured in the lab. A consequence of these virtual processes are virtual particles, which are particles that quickly materialize and disappear into the vacuum before having any material consequences. The vacuum is not empty but seething with rapid interactions of quantum fields and particles. Remember that quantum fields comprise oscillator modes that vibrate in the vacuum state. These field oscillators also interact with other field modes. A virtual process occurs when field modes spontaneously activate to create a particle, and like a fish leaping out of the sea the virtual particle quickly returns back to the vacuum state. An example of this is when an electron and positron get pair produced from field oscillations and almost immediately annihilate into the vacuum state. Therefore under normal circumstances, these virtual particles never materialize into real ones.

It is the uncertainty principle that is at the heart of virtual processes, and it is remarkably dictated by the following relation, which says that the uncertainty in the energy of a virtual process is inversely proportional to the uncertainty in time that the process occurs. According to the uncertainty principle, extremely microscopic time scales correspond to very high energies, enough to pair produce matter and antimatter from the vacuum state. For all practical purposes, the vacuum is continually creating and destroying particles at a rate so fast that we do not "see" them because they are too short-lived.

Using Einstein's relation that equates energy with mass, we can find the condition to create two particles of a given mass out of a field that carries the necessary amount of energy. As an example, if we combine Einstein's energy-mass relation, stated above, along with the uncertainty principle, we find that it is possible to create electrons and positrons during a time interval that is roughly a trillionth of a trillionth of a second! During that time the particle and antiparticle pop into and out of existence (because they soon annihilate). That's why we never see virtual processes in everyday situations. As we will now see, according to inflationary theory, all the matter that exists around us, including ourselves, were originally virtual particles. How is this possible?

In everyday situations where the curvature of space-time is nearly flat, virtual quantum fluctuations are incredibly short-lived and remain quantum—they never take on all the characteristics of a classical particle. But inflation produces these virtual particles, and somehow they do become classical. Recall that in quantum

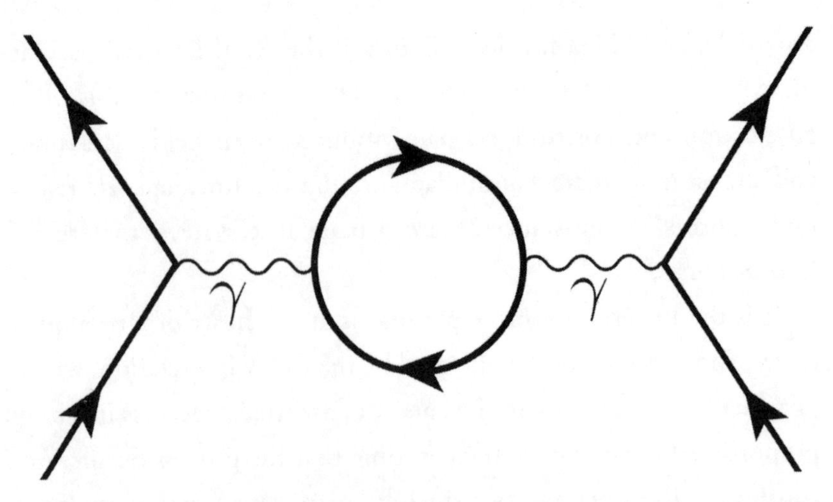

FIGURE 17: A Feynman diagram of a virtual process. The wavy line represents an incoming photon, which emits a virtual particle, represented by the closed loop.

mechanics, the collapse of the wave function requires a mechanism to solve the measurement problem. The Copenhagen interpretation postulates that a classical measuring device or observer collapses the wave function. However, there are no known observers or classical apparatus during or soon after inflation that we can postulate to collapse the wave function of these virtual states. Therefore, inflation has a cosmological measurement problem.

However, due to the dynamic nature of the expanding space-time during inflation, virtual processes play a central role in becoming the very seeds of structure in the universe. Let's take a deeper dive into the virtual processes during inflation. One interesting phenomenon is the exponential expansion of space-time feedback into the dynamics of the quantum vacuum fluctuations, causing them to reduce their uncertainty. The reduction in uncertainty transmutes the oscillations to be a coherent state. This might sound fanciful, but in fact, these states are identical to when photons collectively become a coherent laser beam. These coherent states undergo a phenomenon called

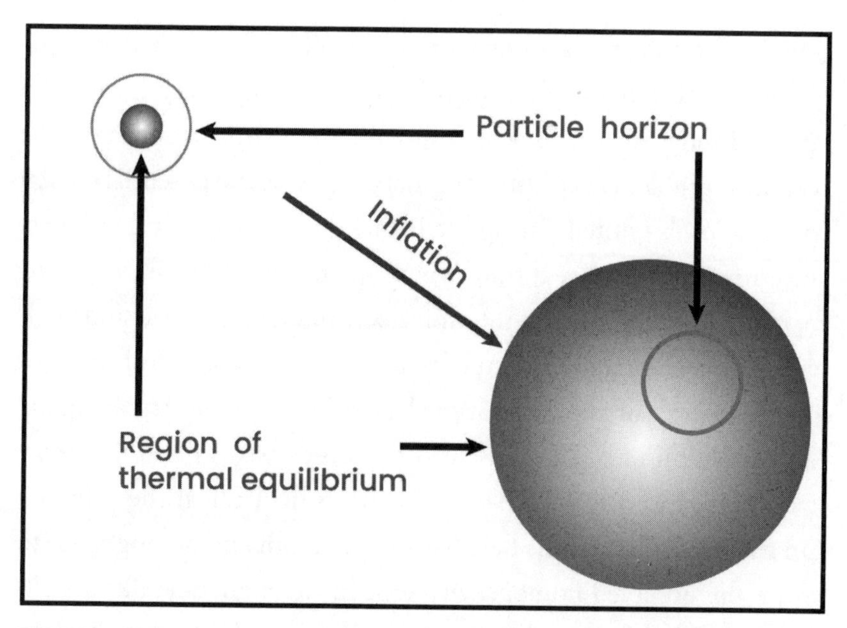

FIGURE 18: Schematic representation of inflation in the universe.

quantum decoherence, which washes out all the quantum phases in the wave function, rendering the primordial quantum fluctuations into classical seeds for cosmic structure. The key to virtual processes during inflation is that the time during which inflation happens is shorter than the time scale, dictated by the uncertainty principle, necessary to produce the observed particles around us. Inflation can last as quick as a millionth of a billionth of a billionth of a second. During this short time, the inflaton dumps its potential energy into the production of virtual quantum particles. These virtual particles live long enough and simultaneously get stretched to transition from virtual particles to real classical particles such as electrons, quarks, photons, and others produced in the big bang.

Cosmic inflation was invented by my mentor Alan Guth to solve a handful of problems that plagued the standard hot big bang paradigm, especially the horizon problem that we discussed earlier. Inflation's magic stems from a short burst of exponential expansion of the universe, transforming a microscopic universe into the macroscopic one we inhabit. Soon after, the inflationary theory was improved by Andrei Linde, Andy Albrecht, and Paul Steinhardt to address some technical problems in Guth's original proposal. The potential energy for inflation is ignited during the Planck epoch, which is the earliest stage of the big bang, during which quantum effects are expected to reign supreme due to the universe's microscopic size and high energy scales. A large amount of potential energy can transform into kinetic energy of the gravitational field, creating an exponentially fast expansion rate. The consensus from theorists is that this energy is contained in a new spin-zero quantum field called the inflaton. On the largest scales, this field has to be smooth and homogenous to create the observed homogeneity seen in the night sky: like a well-shaken bottle of milk, the sky looks pretty much the same whichever direction you choose to look. During the rapid expansion, a few magical things simultaneously happen. First, all present quantum field oscillations get created from the vacuum and stretched by the expansion. This happens because the part of the gravitational field that expands, known as the scale factor, directly couples to the wavelengths of quantum vibrations, acting like a volume amplifier in a cosmic stereo. As the scale factor increases, wavelengths likewise get stretched.

This basic picture of cosmic inflation is compelling and makes the correct prediction of the observed properties that we measure in the fluctuations of the cosmic microwave background radiation. Indeed, the so-called empty space in the early universe, if it is dominated with vacuum energy, comes alive through the engine of inflation, to convert this potential energy into the observed particles and radiation we see. The magic behind inflation is its effect on general relativity to create an exponential expansion of space itself. All that existed in this infant universe were quantum fluctuations riding along with this inflationary era. And it is their particle excitations

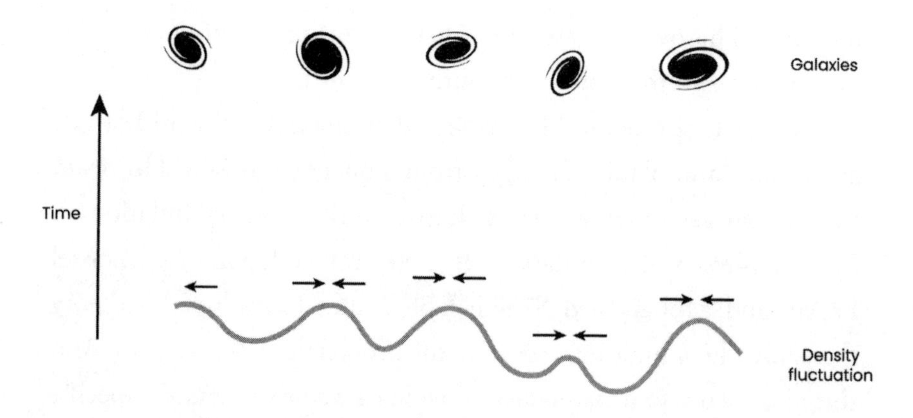

FIGURE 19: The wiggly lines represent the quantum density fluctuations sourced by the epoch of cosmic inflation. At later times these density fluctuations grew, with the help of dark matter, to form galaxies.

that get created out of the vacuum of space-time that becomes all the observed matter around us, including us. However, there remain some foundational problems of inflation that need to be addressed.

The quantum birth of matter and antimatter during inflation is a very promising avenue for generating the observed asymmetry between matter and antimatter that characterizes our universe today—that is, for solving the problem of baryogenesis. Inflation does the job of producing matter, but it also produces antimatter. Unless matter over antimatter can be produced either during inflation or after, we are left with a world of equal matter and antimatter, and that's not our world. If there were a way to bias the production of matter, then we would explain a big part of why we don't see galaxies, planets, and stars made of antimatter. There have been heroic attempts to account for baryogenesis in the early universe, and most models assumed that the necessary ingredients occurred after inflation, and this made good sense. The rapid expansion of inflation expands the volume of space so much that an initially large number of particles would dilute away. But aside from this issue, two conditions spelled out by Sakharov are naturally satisfied during inflation—particle production and the out-of-equilibrium condition.

The key lesson behind inflation is that our classical world of galaxies, stars, and planets emerged from a quantum origin. However, the baryon asymmetry was established either during inflation or by new physics after inflation. In 2004 my collaborators Michael Peskin and Mohammed Sheikh-Jabbari and I developed a theory that shows how inflation can solve the problem of baryogenesis. And this idea, many years later, also provided a way to generate a specific breed of dark matter. The key to this mechanism is to realize that the very agent that ignites inflation, the inflaton field, can interact or couple with matter and antimatter in an asymmetric manner that violates CP and baryon number in one fell swoop. This is elegant because instead of struggling to satisfy the three Sakharov conditions independently, the inflaton meets them all in one shot.

That doesn't solve all the challenges facing inflation, however. Inflation assumes that quantum fields, including the inflaton field itself, are subject both to the laws of quantum mechanics and to gravity. Gravity is mostly classical. However, for inflation to do its magic, some aspects of gravity have to be quantum mechanical as well. The seeds of structure, including the virtual processes, are genuinely born when they interact with quantum undulations of the gravitational field. These undulations are small compared to the part of the gravitational field that undergoes homogeneous growth in three spatial dimensions. But we are free to ask, why not quantize all gravity? Nothing stops the tiny quantum fluctuations of the gravitational field from growing the closer we get to the time that inflation begins. Why should we fully quantize the matter fields and not fully quantize gravity?

The quantum fluctuations of inflation also involve quantum fluctuations of the gravitational field. And as we go back to earlier times when inflation itself was ignited, these quantum gravitational fluctuations approach a big bang singularity hinting at a need for a full treatment of quantum mechanics and gravity, a theory of quantum gravity. We will see that quantum gravity and inflation forces some sharp questions about assumptions we make about physical reality. But to get to the question of gravity, let's consider something like antigravity first.

10

EMBRACING INSTABILITIES

A global virus pandemic brings the world to its knees. The stock market suddenly drops, inciting fear from investors. A star collapses to form a black hole. These events all have one thing in common. They are all instabilities that correspond to a catastrophic growth in some quantity that leads to an unwanted outcome. In physics, sometimes the bad outcome of an instability threatens to obliterate the validity of the theory itself. The quantum revolution was born in part as a result of taming instabilities in atomic systems, such as the ultraviolet catastrophe and the instability of classical atomic orbits. In recent times, billion-dollar particle accelerators were built to look for supersymmetric particles that function to tame an instability that would lead to a catastrophic growth in the masses of all matter. No diet would help us for such an instability. Many physicists were confident that this pattern of fixing instabilities would lead to experimental confirmation of supersymmetry, but it didn't happen. Are all instabilities tragic, or are some useful for our universe's functioning in hidden ways that could lead to new directions in our understanding? What are instabilities, especially of the quantum type, trying to tell us about the new physics?

My colleague João Magueijo at Imperial College developed a theory of the early universe that solved the infamous horizon problem and flatness problem in cosmology. In ordinary relativistic cosmology, the horizon problem describes the fact that the edges of the universe seem to have been in communication with each other, despite being too far apart for light to have traveled from one far-flung region to another in the time the universe has existed. The flatness problem has to do with why there's just the right amount of energy in the universe so that its geometry is Euclidean and flat. In Magueijo's theory, he posited that in the earliest stages of the universe, the speed of light could vary in time, at odds with Einstein's postulate of the constancy of light. Magueijo presented a mathematical model of his theory in a publication that was later contested by some theorists as being a sick theory, because it seemed to give rise to instabilities if the theory were to be quantized—to some aficionados, it was not even wrong. I was there for the showdown, and Magueijo, being a black belt, enjoys these physics sparring sessions. We were sitting at a seminar and the invited speaker made reference to Magueijo's theory and claimed that it was sick because it had an instability. Magueijo responded, "I want the damn instability. After all *you* are an instability!"

What Magueijo meant was that the chain of cosmic events leading to life were ignited by instabilities. These tiny perturbations in the CMB that grew to form the first structures is an instability in the equations that govern those cosmic structures. These good instabilities get regulated and stabilized when the system becomes nonlinear and highly interdependent with the gravitating environment. Today physicists are wrestling with strange instabilities that continue to cause the physics community headaches. We will place our focus on an instability that is found in empty space, a vacuum instability that communicates with everything otherwise known as the cosmological constant or dark energy problem.

Recall that the discovery of quantum mechanics was ignited by a handful of instabilities found in classical physics. Our very existence owes to the stability of the atomic structure of hydrogen and oxygen in water molecules, but classical physics predicts that all electrons should spiral into protons, making the classical atom unstable. Every time the electron orbits around the proton it radiates away electromagnetic energy, which reduces its distance to the proton. Eventually it spirals into the proton, rendering the atomic system unstable. By quantizing the orbits of the electron by associating each orbital distance with a wave, the electron, like the lowest vibration on a guitar string, would have a lowest orbit allowed.

Another type of instability is when a system's energy continues to grow without bound. This is similar to falling down an infinite hill—your kinetic energy would continue to increase until it reaches infinity. And this contradicts relativity, which says nothing can go faster than light. Ironically, while quantum mechanics was invented to tame classical instabilities, it was later discovered that even quantum systems can have instabilities. For instance, when we consider the quantum effects of electrons, what we perceive as empty space in front of us turns out to contain a form of energy due to the activity vibrating quantum fields. This contribution to energy (often called vacuum energy) is very large and should dominate the energy of the universe, causing this expansion to accelerate so fast that galaxies would not have a chance to form. This is known as the cosmological constant or dark energy problem, and as of today there remains no solution. Let's understand this at a deeper level.

As we explored previously, the fundamental substance behind all matter and force carriers known to us are quantum fields, and they all generate vacuum energy. As we saw earlier, one consequence of quantum fields is that empty space is not empty; it is filled with vibrations of quantum fields that, in the universe's past, built all substance from stars and planets to life itself. Thus, empty space is occupied with these vibrating, interacting quantum fields. Recall that a quantum field has properties like a large collection of oscillators. Because of the uncertainty principle, a quantum field oscillator can never be at rest.[1] For such a system the rate of vibration is proportional to the field's energy and can emerge as a particle from empty space like a water droplet that jumps on the surface of the ocean.

The baryogenesis problem makes clear that just as every particle is created, so too is its antiparticle created, but only for a short moment of time, beyond our ability to see with our naked eyes or our most advanced high-speed shutter cameras. This spontaneously created pair of particle and antiparticle quickly meet and annihilate each other, back into empty space. What's the use of that? This process is called a vacuum bubble, and it generates energy. Gravity likes energy. There are other similar effects that likewise generate vacuum energy. In every chunk of empty space, the vacuum energy can be envisaged as completely transparent quantum fluid that differs from ordinary fluid. If you try to compress an ordinary fluid like water, it will resist compression and push back on you; this is positive pressure. The vacuum fluid has negative pressure and does the opposite. Try to compress it, and it will expand outward. The effects of quantum vacuum energy in empty space have been confirmed in the lab and they are tiny. But the predictions of our standard model predict a much larger amount of vacuum energy. And all forms of energy according to general relativity, including vacuum energy, will warp space-time. So, how exactly does vacuum energy affect curved space-time?

If there is one key lesson to take away from Einstein's theory of general relativity it is that space-time is also a field. The matter fields and their vacuum energy are tethered to the gravitational field, causing it to warp. The vacuums' negative pressure stretches space by accelerating observers away from each other, resulting in a repulsive force. For instance, if there was a lot of vacuum energy near

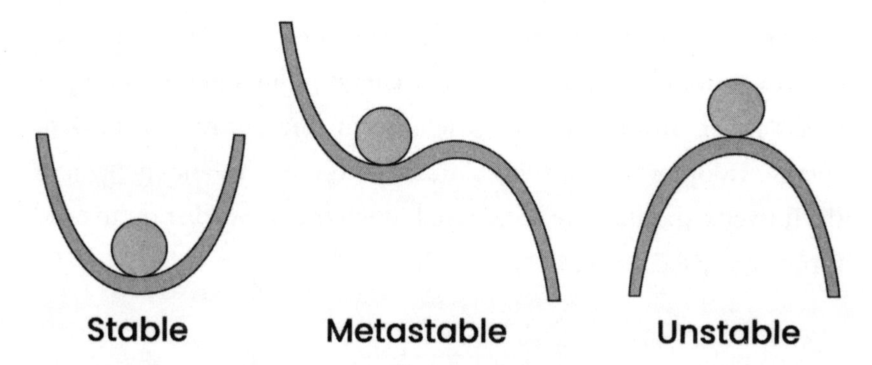

FIGURE 20: Systems can be stable if the potential energy is bounded from below, as seen to the left. The middle metastable graph shows a bounded potential locally, but unbounded to the left. So, this particle can be both stable and unstable depending on its energetics. An unstable system has a potential that is unbounded in all directions.

the earth and the moon, space would stretch rapidly outward and they would fly away from each other. Now comes the punch line. In 1998, groups led by astronomers Saul Perlmutter, Adam Riess, and Brian Schmidt observed with powerful telescopes how distant, exploding stars recede away, and showed that the expansion of the universe is accelerating. The only way this can happen is if the universe were filled with vacuum energy. But there is a major problem: according to precision calculations in quantum field theory and general relativity, the expected and observed vacuum is 120 orders of magnitude smaller than what our standard model predicts. You may wonder why there is all the fuss about this discrepancy between our most precise theoretical characterization of modern physics and our observations. For one, it says that something is wrong with either general relativity or quantum field theory, or both, in describing the quantum effects of fields on space-time. Alternatively, there is some unknown reason why the vacuum energy is either hiding itself or tamed by some new, mysterious, unknown physical force—yet to be discovered.

The form of vacuum energy that generates the universe's acceleration is popularly known as dark energy. Our understanding of precision physics predicts an instability in the production of dark energy. In other words, there should be too much dark energy and the universe should have accelerated much too fast earlier on to even form stars, galaxies, and us.

After a research stint in London, I headed out to the hills of Silicon Valley to continue my research at the Stanford Linear Accelerator Center (SLAC), known as one of the meccas of particle physics. My new colleagues at SLAC and Stanford were some of the pioneers in string theory, a theory that realizes Einstein's dream to unify all four forces. The idea of string theory is simple and elegant; the articulation of the theory requires some of the most advanced mathematics known to us. My new colleagues had worked for decades along with a global community of string theorists to develop this elegant ten-dimensional, yet abstract framework. Young physicists like myself were handed the baton in this quest. String theory provided a rich framework to address problems that share residence in particle physics and cosmology. This is mainly because string theory naturally links the extra dimensions with gravity and hidden symmetry patterns found in the standard model of particle physics.

There are strong reasons to believe that string theory could very well be the theory that underlies our reality. As we will explore in an upcoming chapter, string theory unites quantum mechanics with general relativity by taming disastrous ultraviolet (short wavelength) quantum instabilities.[2]

Once cosmologists measured the amount of dark energy (also known as vacuum energy or the cosmological constant) deduced from the acceleration of space-time, we had to reckon this fact with our precision theories. We know that the standard quantum field theory interacting with general relativity was not enough. This was an opportunity for more fundamental theories such as string theory, which had ingredients of taming certain short scale instabilities, to come to the rescue. We saw this as some experimental guidance for our theories, places where the theories could possibly make contact with the real world and tell us why the universe was accelerating at this epoch in cosmic history. However, a major set of challenges presented itself. String theory is blessed with symmetries that tame infinities, but the symmetries themselves did not straightforwardly allow for a positive cosmological constant, let alone the tiny value that exists in our universe. I approached the head of our group and asked him for some guidance in how I and the other postdocs might navigate through the difficulties and confusion. His advice was to find an example of a solution from string theory that admits any positive cosmological constant. We were all on a hunt to find this solution. Within a year my colleagues Shamit Kachru, Renata Kallosh, Andrei Linde, and Sandip Trivedi found a remarkable pathway and solution of realizing a cosmological constant in string theory. And this had to do with finding a special space-time solution in string theory called de Sitter space.

Solutions of general relativity that admit a positive cosmological constant were known since the inception of the theory. One solution was discovered by Willem de Sitter, and as expected, it gave accelerating space-times; the larger the cosmological constant, the more the space accelerates, and by an exponential amount. During the early stages of the universe's expansion, there was a fine balance between the repulsive expansion of space and the attractive gravitational pull of infalling matter to form the first stars and galaxies. The Nobel laureate Steven Weinberg realized that if there was too much vacuum energy in the universe, the repulsive force would overshadow the attraction necessary to form structures and there would not be any structure, hence there would be no life in the universe; there would be no universe as we know it. The universe seems to have dialed in just the right amount of vacuum energy for structures and life as we know it to exist. So it is up to our theories to understand not only why vacuum energy is positive but also why the exact amount needed for the existence of the universe is observed—we call this the "why now?" or coincidence problem. Unfortunately, as we've seen, our current understanding of physics produces far too much vacuum energy. So, the real question is: What happened to all the vacuum energy that we expect to exist? Or let us assume that all this vacuum energy does exist, so then why does gravity not respond to vacuum energy? This remains a mystery.

During my time at Stanford, young string theorists like me were seeking out pathways in the jungle of ten-dimensional calculations to find solutions that have a small and positive cosmological constant. Then, in a groundbreaking paper, renowned theorists and friends Raphael Bousso and Joe Polchinski argued that the search to find such solutions could be futile. I'll come back to their work later in the book; for now it's enough to say that it essentially argued that the universe was just one of many, which have a vast range of possible values for the cosmological constant, without any first principles to force its value. By reasoning that has come to be known as the anthropic principle, they argued further that we necessarily live in a universe with a cosmological constant that is capable of supporting life like us, and so we shouldn't seek any deeper explanation for it. A big debate transpired in the cosmology and string theory community as to whether the anthropic principle was scientific. I decided to take another direction, which would risk further shunning from my colleagues. This new direction would mean that I would engage in conversations with the outsiders from my club and even import their ideas into a possible resolution of the cosmological constant problem. I was guided by the motto: "Let the nature of the problem dictate the tools you should resort to," even if it meant borrowing from the outsiders or risk becoming one yourself. Active deviance was on the horizon.

That I even wanted to work on the cosmological constant problem was already deviant behavior. In the aftermath of that paper, postdocs were warned to not work on it. Some of my advisers suggested that we postpone working on the cosmological constant until we got tenure. It was a Medusa that defeated the morale of many theorists who attempted to decipher one of nature's biggest puzzles. But I was haunted by the beauty of the cosmological constant problem and was fine with joining the ranks of those that it defeated. Plus, it was my last year on the job market for a faculty job, and I did not have high expectations of getting a permanent position, so I did not feel like I had much to lose.

If there was any chance to tread forward, I would have to find a new direction that was not thought of before. One day while having coffee on top of a hill in Nob Hill, I saw a similarity between the cosmological constant problem and a problem that haunted particle physics, the strong CP problem that came up in baryogenesis and the origin of our matter-filled universe. In the strong nuclear interaction, described by a theory otherwise known as quantum chromodynamics (QCD), the gluon is the particle that mediates the interaction between quarks that bind to neutrons and protons. Classically the neutron is electrically neutral, but quantum effects induced by QCD create a very large amount of net electric charge in the neutron that would destroy the stability of atoms. This would be catastrophic, given that we're made up of neutrons and protons. Like the cosmological constant in general relativity controlling whether the universe expands, is stable, or contracts, there is a parameter in QCD, called the theta parameter, that controls the amount that the neutron deviates from electrical neutrality. Experimentally the theta parameter was measured to be on the order of one billionth.

So, the strong CP problem is relegated to a question as to why the theta parameter was so close to zero, a similar predicament as the cosmological constant.

In the late seventies, Helen Quinn and Roberto Peccei found an elegant solution to the problem. They realized that the strong interactions could have a hidden new symmetry. Think of this symmetry like a particle rotating around a frictionless surface in a perfect circular orbit. If all forces were absent, every point of the ball's orbit would have the same energy. So, the energy has a symmetry such that any rotation of the ball leaves the energy invariant. Now if we add gravity to the problem and slightly tilt the circle, then the gravitational force would break the symmetry, simply because there is potential energy in the gravitational force. Similarly, Peccei and Quinn showed that a quantum effect—the emergence of a new field called the axion—breaks CP symmetry, introducing a potential energy. This axion readjusts itself to the minimum of the potential and conspires to drive the theta parameter to zero. I wondered if we could reimagine the cosmological constant to act like the theta parameter of gravity. Luckily, Helen Quinn was at SLAC. I told her the idea once over lunch.

We sat at a work desk in her office and Quinn meticulously walked me through the conceptual and mathematical inner workings of her solution to the strong CP problem. There is a big difference between reading a research paper on a physics result and learning it directly from the author herself. It was only when I saw how Quinn thought about the strong CP problem and her and Peccei's ingenious insight into the solution that I was able to continue to find a way to implement the idea into gravity and the cosmological constant problem. Most importantly, while Quinn had high standards for the creative and technical implementation of the idea, she was encouraging, and it empowered me to take the next step.

To take the analogy between a problem in QCD and gravity to a place where I could attempt to do a calculation, it would help if I could place gravity on similar footing with QCD, and there was one activity of research that already did that. In the arena of quantum gravity, string theory is seen as the only game in town, but it isn't. There are other attempts to quantizing gravity, even if they do not sit well with my string theory friends.

One particular approach is loop quantum gravity (LQG), in which the starting point is to quantize gravity using the same methods as QCD. This possibility came from Abhay Ashtekar's ingenious insight to rewrite general relativity using variables identical to QCD.[3] The idea required me to use the Ashtekar formulation of gravity in the presence of a cosmological constant. When I spoke to the other postdocs about loop quantum gravity, many of them dismissed the theory as "loopy" and suggested that anyone that would work on that theory does not know physics. But I already felt like an outsider and pursued loop quantum gravity anyway. Besides, when I pressed some members of the group to provide a solid critique about loop quantum gravity rather than just make fun of it, most of them did not know the theory well enough to tell me why they thought it was wrong. So, I decided to invite one of the founders of loop quantum gravity, Lee Smolin, to come to Stanford and SLAC to give some lectures on the theory. That way at least we could have a more informed critique of the theory as a group, and I could make progress on my idea for the cosmological constant. I finally learned enough about the Ashtekar formulation to get going on the project. But my invitation to let an outsider come into our club to teach us something left a bad taste in the mouths of members of my group.

That was just one of the reasons I spent much of my time at a café called Cup of Joe near the top of Nob Hill performing calculations on the project. I would frequently drop by the office of my adviser, Michael Peskin, to discuss my work when I got stuck. Finally, I finished the project and provided a mechanism to partially solve the cosmological constant problem in a manner similar to the

Peccei-Quinn mechanism. The key idea is that there is a new axion that, like a person wobbling on a high wire, can readjust itself to cancel the cosmological constant. The model still needed more improvement to address the other aspects of the cosmological constant problem. However, my results were enough to put the paper up for publication, and the idea inspired a new way to think about resolving the disastrous production of dark energy.

The paper, entitled "A Quantum Gravitational Relaxation of the Cosmological Constant," was posted on ArXiv.org, a website where physicists share drafts of their papers with the global physics community. Days of silence from the community went by. I was not offended. My office mate, string theorist Amir Kashani-Poor, returned after giving a seminar at the University of Texas at Austin. He told me with a look of awe that he had been lunching with the string theorists, and Steven Weinberg had joined the group. Weinberg shared the Nobel Prize with Abdus Salam and Sheldon Glashow for unifying the electromagnetic force with the weak interaction and is known as a straight shooter. Kashani-Poor told me that Weinberg pulled my paper from his inside sport jacket pocket and said to the group, "Have any of you seen this paper? It looks really interesting." This was especially vindicating since Weinberg wrote a seminal masterpiece on the various problems with the cosmological constant and the attempts to solve it. Weinberg's work also provided concrete criticisms and no-go theorems that ruled out many attempts to solve the cosmological constant problem. Luckily my model was able to evade Weinberg's no-go theorem. My model is still a work in progress, and my research group and I are still improving it to confront how the cosmological constant is related to the onset of dark energy today.[4]

This was my first step toward really working on quantum gravity. Aside from the sort of aesthetic question I raised earlier in the book about the big bang—why should gravity only be partially quantized?—there are also pragmatic scientific reasons to seek a quantum theory of gravity. Currently, there are a handful of unsolved issues at the interface of quantum field theory and classical gravity. And many of these problems find a home in the beginning and early evolution of the universe when both gravitational and quantum physics are expected to be active. These cosmological problems, such as the origin of matter over antimatter, inflationary versus cyclic universes, dark matter and dark energy, are so daunting that they compelled me to deviate again and carry out research in both superstring theory and loop quantum gravity. My collaborators and I have spent the last two decades using these cosmic conundrums as a compass to test and further develop theories of quantum gravity.

Both LQG and string theory have their own communities, and there are strong feelings about the veracity of each approach. Both have complementary strengths and weaknesses, ranging from technical to conceptual challenges. My take is that both theories provide tools and new concepts to address the unresolved problems facing cosmology and particle physics, and both may not be sufficient. It has always been my take to let the problems and unexplained observations dictate which theories and tools to resort to. Thus, I have had the fortune to use both LQG and string theory in my research. While we have a handful of candidates for quantum gravity, none are complete. It is my view that they are all parts of the elephant, and we can use these approaches to teach us something about what the final theory might look like.

11

A COSMOLOGIST'S VIEW OF A QUANTUM ELEPHANT

We don't have to get to some altered state of mind or enter a wormhole to see how mysterious the universe is—it's right in front of our faces. Our electronic devices are controlled by quantum effects of electrons, and yet, we still don't have a clue about what the ghostly electron is doing as it approaches the two open holes in the double slit experiment. It does something, but whatever it does, it seems like our mind's eye is unable to visualize or conceptualize it. The uncomfortable picture is that the electron behaves like a wave when we are not looking at it, and not just a material wave but a wave of potentiality—it's as if some invisible puppet master is playing tricks on us. If we try to interrogate the electron at the two open slits, where we expect it to be doing its witchcraft, it goes back to behaving like a particle, and the interference pattern is lost. Another equally weird situation is when two spinning electrons are entangled and separated at far distances. We still don't know how one electron can instantaneously know that the other's spin was measured to always have the opposite spin of its distant partner. Quantum theory predicts this spooky action at a distance but refuses to tell us *how* it happens. In other words, the math says that the spins should know about each other at infinite distances, but the math doesn't simply provide a mechanism. These unresolved issues in quantum foundations have led some physicists, including Einstein himself, to believe that quantum mechanics is incomplete. However, as far as precision experiments can tell, all the predictions of quantum mechanics, no matter how bizarre, have been confirmed.

To me what is bizarre is how this real world, with planets, people, and falling apples, emerged from an underlying virtual quantum universe in the distant past. As I will soon discuss, we have strong evidence that our entire universe in its infancy was quantum, and it evolved to generate our macroscopic world. The paradox can be simply stated as: What measured the wave function of the universe? A first step to understanding the magical emergence of our classical world is to ask how quantum mechanics and Einstein's description of gravity and space-time can be merged.

We discovered that when the gauge invariance principle is united with special relativity and quantum mechanics, we have a pathway to unify three of the four forces. We arrive at a unique quantum field theory called the standard model that puts all three forces on the same conceptual and mathematical footing. For decades physicists have been challenged in uniting gravity with quantum mechanics. Here is the elephant in the room: it is space-time that would be the state that will exhibit superposition. So, what does it mean for a region of space-time to have many superpositions? Despite this and a handful of potential conceptual questions, there is good reason to invite the quantum into space-time, since quantum mechanics has a history of resolving the singularities and instabilities found in classical physics. Classical general relativity, which is home to even worse black hole and cosmic singularities, seems to be begging for a quantum rescue. What is the culprit that makes quantum gravity so difficult? We must go into this abyss as we expect the final theory of quantum gravity, if it exists, to see what's behind the black hole singularity and what happened before and at the birth of our universe. For me, quantum gravity holds the promise to give us a new view of our universe far beyond our current imaginations. Maybe we will be able to make sense of the hidden variables that Einstein was after.

First, it's useful to understand the two main doorways into quantum gravity: background independence and emergence. Background independence is a strong version of the principle of invariance. Recall that the principle of invariance is a key concept in all four forces, but it has a special role in general relativity: invariance operates differently in gravity. The differences give us the first key as to why quantum gravity differs from the successful quantization of gauge field theories like the standard model.

We use coordinate systems to describe the position and motion of an object, but according to the equivalence principle, the reality of the object should not depend on the coordinate label given. We could use any other coordinate system to describe the same particle. Einstein realized that the laws of physics should not depend on the choice of a coordinate system—this is the principle of general covariance, an invariance principle for coordinate systems. The invariance of coordinates is analogous to the name given to a person. Suppose there is a person whose name is John. In Spanish, his name is Juan. In Portuguese it is João. Regardless of the various translations of the name, they are just labels for the actual person, who is independent of the translations. So, if we can describe the location of things with a coordinate system, then different "translations" or choices of these coordinates are all equivalent ways of describing the location of the thing. In principle, there is an invariance encoding the infinite number of ways we could label the position of an object at a point in space-time. We call this diffeomorphism invariance, and it is at the heart of the relativity principle.

Loop quantum gravity is a successful background-independent formulation of quantum gravity. It was pioneered by Abhay Ashtekar, Carlo Rovelli, and Lee Smolin, whose trick for quantizing gravity was to reformulate general relativity so that the background independence is formulated as a gauge theory (like the standard model of particle physics) and to implement similar gauge theory methods of quantizing as was done in the strong interactions of quarks—that is, in quantum chromodynamics. A gauge field can be quantized by having it wrap around a closed loop, like a lasso. The wrapping of a concentrated amount of gauge field flux around a loop is known as a Wilson loop, named after its inventor, Nobel laureate Kenneth Wilson, and it was used to understand how quarks confine in the strong interactions. We can think of Wilson loops as tubes that carry electric flux—lines of a concentrated electric field. The idea is that two quarks are attached to each other by a string of flux. Similarly, gravity is equipped with its own gauge field called the Ashtekar connection. This connection is the key ingredient behind quantizing gravity, and it leads to the remarkable conclusion that the basic building blocks of LQG are atoms of space-time called spin networks, invented by Roger Penrose. These networks join to make quanta of area and volume that have the size of the Planck length—a minuscule distance defined by quantum theory as roughly .00000 000000000000000000000000000016 meters. If this is the case, then it points to a way to tame the ultraviolet infinities found in quantum field theory since quantized space has a minimal distance and energy.

Another interesting feature of spin networks is the quantum nature of time. When a spin network undergoes a quantum transition it corresponds to a change in time, which ticks as a quantum by a billionth of a trillionth of a trillionth of a trillionth of a second!

Since LQG is a fundamentally discrete structure, one of the big questions is how to recover the smooth space-time that we see in our

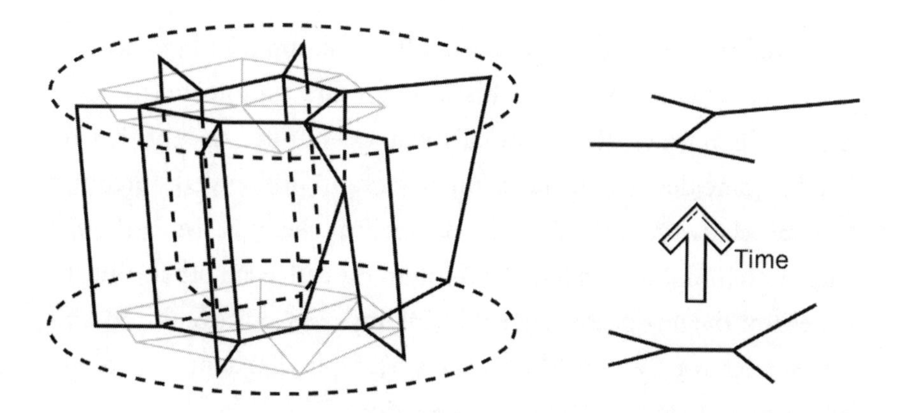

FIGURE 21: A spin-foam transition from one spin network to another depicting quantized time steps.

macroscopic world—this is called the semiclassical limit. In ordinary quantum mechanics the Planck constant is the parameter that turns quantum effects on or off. So, by setting the Planck constant to zero we expect to obtain classical physics. However, this logic does not work in LQG, so some other scheme is needed to get back classical space-time. There have been a handful of proposals to get back classical general relativity from LQG, but it is still at the level of current research. I think that if LQG is to be a successful theory of quantum gravity it should not only tame the black hole and cosmological singularities and recover the good predictions that we already see in general relativity but also connect with some unsolved problems in particle physics.

One question my colleagues and I have been pursuing is how to connect the classical limit of quantum gravity to be coincident with particle physics. One exciting prospect is the realization that the Ashtekar variable, a key ingredient in allowing classical gravity to become LQG, enjoys a broken symmetry that is identical to the weak force. Is this a coincidence? As I was reminded by my mentor, Nobel laureate Leon Cooper, "If we knew what we were talking about, we wouldn't call it research." I vividly remember the moment

I realized this coincidence between the weak force and gravity as a young faculty member at the Institute for Gravitation and the Cosmos at Penn State. It was actually a simpleminded insight that related a coincidence that the symmetry group of the weak force and general relativity, in the Ashtekar formalism, were the same, though these two forces should have nothing to do with each other. But the genesis of the intrigue occurred during my days at the Stanford Linear Accelerator Center; I had always been puzzled about the origin of parity violation in the weak interaction.

Parity, as we saw when discussing baryogenesis, is a symmetry transformation that is analogous to looking in a mirror. When you look at your left hand in the mirror it is identical but gets converted to a right hand. Parity is a transformation that also inverts an object, turning it upside down. For example, parity would transform a spinning ice skater into an upside-down ice skater spinning in the opposite direction. All the forces were thought to be invariant under parity transformations. In other words, if we look at the scattering of a left-handed quark with a left-spinning gluon field it would occur at the same probability as a right-handed quark interacting with a right-spinning gluon field. The big surprise was that the weak interaction allowed interactions with one-handedness.

While it seems that the violation of parity is inconsequential, life would not exist without it. The parity violation in the weak interaction is at the heart of all nuclear decay processes as well as the chain of subnuclear reactions necessary for fusion. The sun relies on the nuclear fusion of two hydrogen atoms to make helium to provide the energy necessary for life on Earth. So, there would be no life as we know it without parity violation. Sheldon Glashow, Abdus Salam, and Steven Weinberg were awarded the Nobel Prize for constructing the correct theory of the weak interaction and uniting it with the electromagnetic force. However, in their model of the weak force the parity violation was assumed, and there was no deep reason as to why parity violation emerged from some deeper theory and how it would do so. My supposition from the resemblance of the Ashtekar connection of general relativity was that parity violation in the weak interaction came from quantum gravity itself.

So, what does one do when an idea appears that seems crazy and too speculative? Well at the time, my strategy was to go to the source, Abhay Ashtekar, the inventor of the Ashtekar variable.[1] Ashtekar has a reputation for being an extremely technical and detail-oriented kind of theoretical physicist, which he combines with laser-sharp intuition. So, if you have an idea and take it to Ashtekar, expect to walk out with your idea in a coffin to be buried in mathematical soil. I knew of young physicists who would be very careful to iron out any technical details before talking with the master. But for some reason, Ashtekar always gave me a hall pass. He treated me as if I had something to say even though my ideas were often raw and inchoate. So, I felt comfortable telling him the idea. After a few minutes, he told me that he didn't notice this connection between parity violation in gravity and the weak force and encouraged and challenged me to pursue the idea. After months of calculations and thought experiments, I developed a theoretical model to explain the weak interactions parity violation as a gravitational phenomenon.

One of the assumptions behind LQG is that the quantization of gravity does not immediately take matter into account, especially fermions. Recall that all known matter fields are fermions, and they are special because they obey the Pauli exclusion principle. Fermions have a geometric aspect to them since they also carry information about the twistiness of space-time, a quantity called torsion. In present versions of LQG, torsion is assumed to be negligible—what physicists call vanishing. But if LQG is to properly speak about fermions then it may have to relax the assumption of vanishing torsion. Fermions are special because the reason your hand does not go through a table is due to the exclusion principle, which states that no

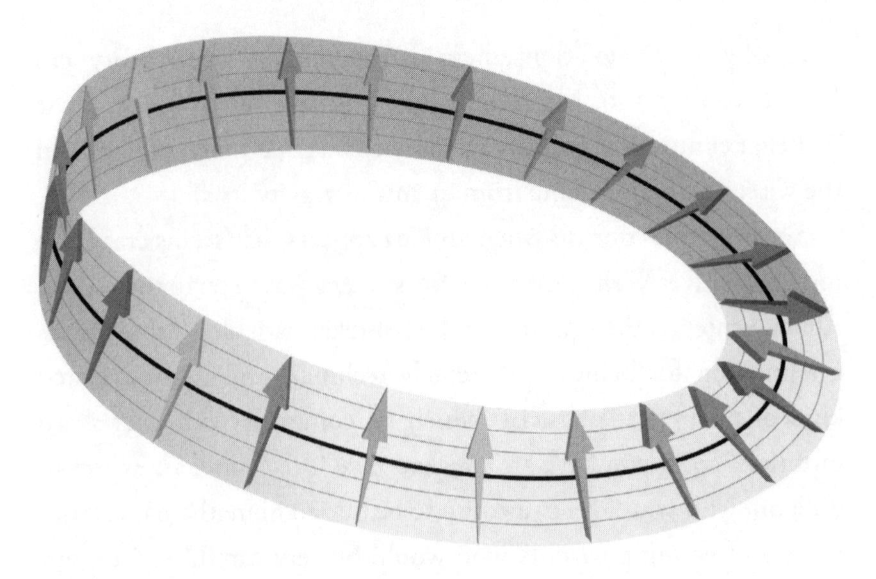

FIGURE 22: A representation of torsion in loop quantum gravity.

two fermions can occupy the same quantum state. So, macroscopic objects exist because fermions, like protons, neutrons, and electrons, completely evade occupying the same place. I believe that if LQG is to have a hope of having a semiclassical limit, it will come from the simultaneous emergence of the exclusion principle from a quantum foam where fermions are not exclusionary. In other words, space-time becomes fuzzy when fermions lose their special exclusionary powers, but some new, as yet to be discovered, physics makes the fermions transition from being nonexclusionary—"fuzzy"—to occupying different points in space-time, linking the fermions' distribution to the emergence of classical space-time.

As I said, LQG isn't the only way that I have tried to understand the elephant of quantum gravity. Recall that our access to the quantum world starts from first considering a classical system and applying quantization to it. As in the case of the standard model and LQG, we can quantize by writing down the relevant equations of the classical system and employing the rules of quantization given by Paul Dirac and Richard Feynman. In the case of string theory, which was first proposed to explain certain patterns in the strong interactions, a remarkable surprise presented itself. Take the description of a string of energy that sweeps out a two-dimensional sheet in space-time. We write a mathematical description of the energy for this string, and we can quantize it the same way we quantize a particle. Naturally, the quantum vibrations of the string come with a richer structure than its point-particle cousin. Like a point particle the quantum string will come in a superposition of vibrational states. However, the string will also have other types of oscillations that are unique to its extended nature.

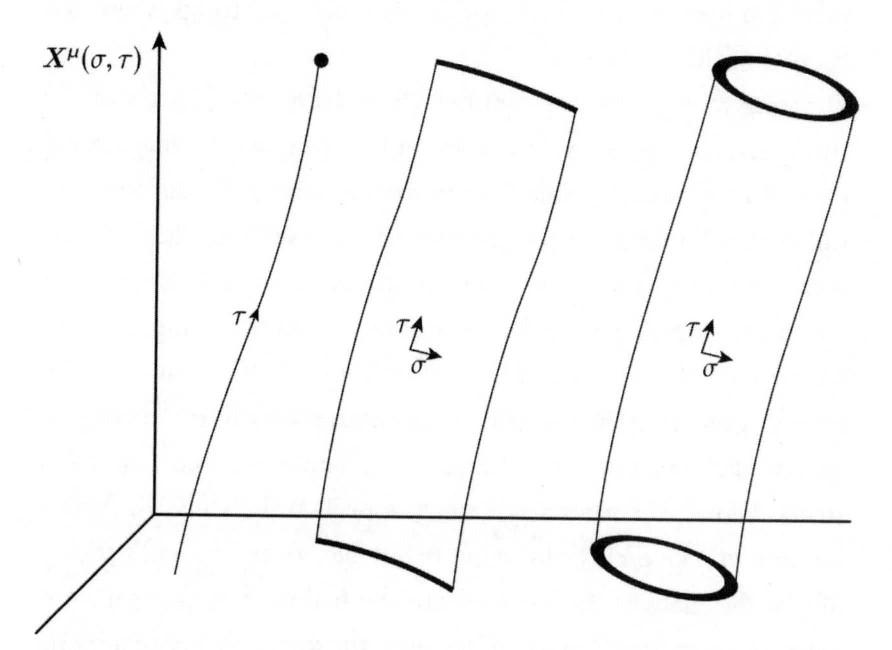

FIGURE 23: Three representations of the key entities in string theory. To the left, we have a point particle sweeping out a world line. In the middle, we have a world sheet of an open string. To the right, we have a world sheet of a closed string.

There are two types of string oscillations, massless and massive. The big surprise to the quantum gravity researchers was that the massless string oscillation includes a spin-2 particle. The only known massless particle with spin 2 is the quanta of the gravitational field, otherwise known as the graviton. This was the first indication that string theory is not what the early string theorist ordered—a theory of the strong interactions. String theory contains quantum gravity.

Even more remarkable is another stringy quantum effect. When a string moves, it can emit a string that propagates and rejoins the original string. This is known as a loop effect, which has a generic effect in quantum field theories to change the strength of interactions. In quantum electrodynamics, a loop effect changes the strength of the interaction between electrons, and this effect has been measured in the lab. In string theory, the same thing happens, but, unlike particle field theories, an invariance that is unique to strings is violated because of the loop effect.

String theory enjoys a symmetry that leaves the physics of the string invariant when we zoom in on the string with a magnifying glass. This symmetry, called conformal invariance, is analogous to the image of a fractal, a pattern that repeats itself regardless of how many times one zooms in or out of the pattern. This symmetry is sacred for strings and must remain intact when the string becomes quantized, otherwise the sacred conservation of probability is destroyed, rendering the underlying quantum theory to make nonsensical predictions. But when the quantum loop effect is considered in string theory, this scaling symmetry appears to be violated. Astonishingly, the loop effect in string theory generates Einstein's theory of general relativity so as to restore the hallowed conformal symmetry of the string. Amazing! We start out with a theory of a string with no gravity, turn on quantum effects, and general relativity pops out. Master string theorist Edward Witten once said that if we knew nothing about general relativity and knew string theory, general relativity would be a prediction.

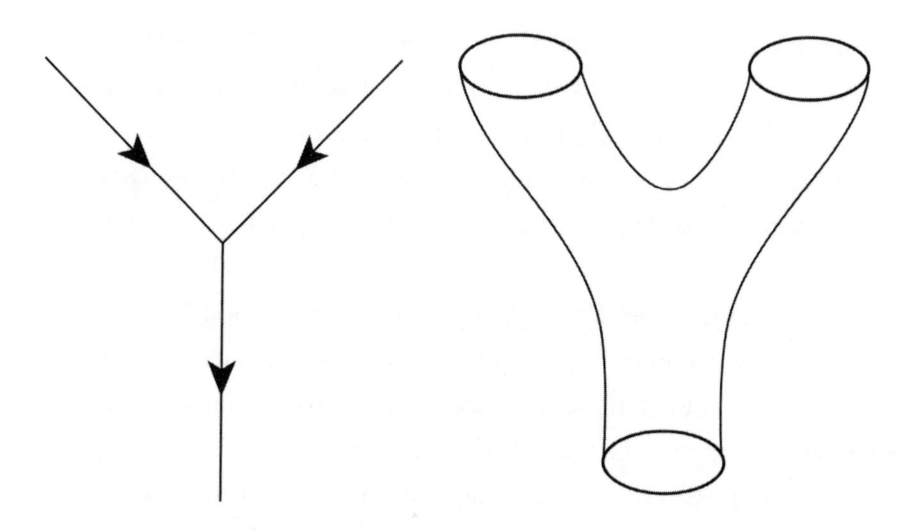

FIGURE 24: Schematic representation of gravity emerging from a string.

For string theory to be quantum mechanically consistent, the resulting theory of gravity must live in ten dimensions. Many physicists have lamented that we live in four dimensions, therefore string theory makes the wrong prediction about the dimension of space-time. However, the extra dimensions in string theory address another mystery found in general relativity and the standard model: What is the origin of the matter fields in our universe? One of Einstein's final efforts for a unified field theory was to extend the notion of geometry of space-time for matter fields. He adopted an idea first developed by Theodor Kaluza and Oskar Klein, who showed in the 1920s that a five-dimensional theory of gravity seen from a four-dimensional perspective is equivalent to Einstein's theory of gravity and electromagnetism. The electromagnetic field is realized as the fifth dimension curled up in a small circle. How does that work? The gauge field, whose quanta is the photon, contains all the information about electromagnetism and arises from an invariance of the photon moving on a circle, which is identified with the symmetry of the extradimensional circle. This is a wild idea—that space can cast itself as electromagnetic fields like a shadow in lower dimensions. Of course, the question of what sets the size and stabilizes the radius of this fifth dimension remains open, and we will return to this issue when we discuss the role of string theory in the early universe. Nevertheless, using the same logic as Kaluza and Klein, we see that the ten-dimensional gravity theory in string theory tells a similar story. The extra six dimensions will carry the information about both the matter and the three force carriers of the standard model.

Because common sense and experimental evidence so far both point to a four-dimensional world, string theory must somehow provide a way to understand why we cannot move through the extra six dimensions and why they are not visible. One criterion is that the extra dimensions are microscopically small. This condition leads to a rich set of predictions. Like the Kaluza-Klein theory, the geometry of the extra dimensions manifests itself as the fields of the standard model. String theory starts to look like both a theory of quantum gravity and a unified field theory. But there are some challenges that have faced string theory for decades. First, on top of the fields of the standard models, string theory also predicts a very large number of other fields called moduli that have not been observed. These moduli are fields that govern the sizes and shapes of the extra dimensions that we do not observe in our low-energy world. String theory comes very close to giving us the field contents and symmetries of the standard model. Recall that the standard model did not give us any information about the coupling constants and masses of the particles. Fortunately, string theory moduli can set the values of these parameters. Since these moduli are dynamical fields in string theory, then what kind of physical mechanism fixes them to the values that are observed in the standard model? If we are to believe that string theory is the answer to reality, then what became of these uncountably large moduli fields?

What I personally find very promising about string theory, despite its issues in giving us a realization of our four-dimensional world, is that it naturally encodes the emergence principle. As we saw earlier, we start with a theory of strings with no gravity and it spits out the equations of general relativity and gauge theories (albeit in ten dimensions) when the string experiences quantum corrections. Also, there are realizations in string theory (and loop quantum gravity) from which gravitational theories emerge holographically. Holography, invented by Gerard 't Hooft and Leonard Susskind, is the idea that physics is encoded in a space that is one dimension less than what is experienced. According to the holographic principle, our experience of this four-dimensional world is actually an illusion, and the information is encoded in a three-dimensional screen that lives at the boundary of the four-dimensional world. A good geometric realization of this is to imagine that our world is a three-dimensional, solid ball. Then by holography all the information about the ball is actually encoded in the two-dimensional surface of the ball, not in its volume. Exactly how the three-dimensional world is reconstructed depends on the exact nature of the model at hand. Hints of the holographic principle were inspired by laws of black hole mechanics discovered by Jacob Bekenstein and Stephen Hawking. They found that for a black hole the entropy did not depend on the volume within the black hole but the area of the horizon. The information of the relevant degrees of freedom was found to be residing on the two-dimensional surface of the black hole and did not care about its interior.

String theory realizes holography by using the correspondence principle that Niels Bohr argued for in quantum mechanics. In string theory there are two theories that are complementary, or dual, to each other. One theory is a nongravitational conformal field theory (CFT), actually a theory of gauge fields, like the ones found in the standard model but with a lot more symmetries. This gauge theory is complementary to a theory of gravity with maximal symmetry with a negative cosmological constant, known as AdS. Since its inception the duality has passed every nontrivial test of the equivalence between these two theories. Taken literally, this means that gravity emerges from the dynamics of the CFT—which itself has no gravity. The string theoretic realization of holography was invented by the Argentinian physicist Juan Maldacena. I remember when Maldacena, at the time a postdoc at Princeton, wrote this first groundbreaking paper in 1997 now referred to as the AdS/CFT correspondence. The AdS/CFT duality relates a gravitational theory to a space-time called AdS, which is a homogenous and isotropic space-time with a negative cosmological constant. I was a graduate student and wondered how he set up the string theory calculation to realize the holographic duality. The paper quickly got international acclaim, and up to today its impact and number of citations by others have not lost steam.

Whatever quantum gravity will look like, I'm convinced that holography will play a key role.

12

THE COSMIC BIOSPHERE

The total number of minds in the universe is one.

—Erwin Schrödinger

I have always wanted to know the answer to three questions: Why is the universe expanding? Is there life elsewhere in the universe? And if so, how did it originate? These might seem like distinct concerns, but I have come to the opinion that the questions about the expanding universe and life are linked. The last several chapters have focused on the first question. Now I want to deal with the second.

The thing is, I don't think they are first and second questions. While we don't have complete answers to these questions, I think origination and persistence of living systems are traced back to the expansion of the universe. You should be at least astonished by this claim, so let's dive into it.

When we look at cosmic history, we see the chain of events that unfolded to generate the large-scale structure in the universe. Cosmic time unfolded with primordial quantum vibrations that blossomed into a hierarchy of stellar environments called galaxies and planets that are now part of an interwoven cosmic structure. When we look at the large-scale structure of galaxies, we assume that life and the universe that it inhabits, borrowing from the famed biologist Stephen Jay Gould, are "non-overlapping magisteria." To most cosmologists, complex systems like life are of little consequence to the problems we are trying to solve, such as the big bang singularity and the parameters of the standard model. To my biologist friends, life is housed in a biosphere that is decoupled from the happenings of the universe out there. But what evidence do we have that life and the universe are truly decoupled? Conversely, what evidence do we have that life and the universe are coupled?

My dance with biology and physics started way back in college, when I considered majoring in biology. I have never been able to shake the biology bug, and my current research in cosmology has reawakened my biological questions, even though I put formal plans aside for a while. At my college, bio majors were required to have a year of organic chemistry, so laziness got the best of me. Still, my musings in biology persisted throughout my studies in physics. In the middle of my second year of graduate school I decided that I wanted a change, so I approached my field theory professor Gerry Guralnik, a codiscoverer of the Higgs boson, and confided to him that I was interested in biophysics and was considering leaving physics altogether. Then Guralnik said, "Let me call my former PhD adviser Wally now and get his advice."

The Wally that Gerry was referring to is Walter Gilbert, a theoretical particle physicist who had also caught the biology bug and ended up winning the Nobel Prize for a key discovery in genetics—the nature of stretches of DNA known as introns and exons—which, among other things, led to the human genome project and gene therapy. Guralnik made the call and within minutes Gilbert invited me to visit his lab at Harvard. I spent three hours talking with Gilbert and he gave me an in-depth tour of his lab. He also made a recommendation for me to work in Harvard's biophysics program. That summer, thanks to Gilbert's recommendation, the renowned biophysicist Jim Hogle offered me a job in atomic resolution virus structure determination using X-ray crystallography. From that experience, I learned some new tools and gained a deep appreciation of the complexities involved in biology and the inapplicability of physical reductionism in attempting to comprehend life's processes. It also comforted me to learn that I wasn't alone: other physicists and mathematicians—including many of the greatest—had explored their own biological musings. People including John von Neumann, Eugene Wigner, Claude Shannon, Norbert Wiener, and Roger Penrose, to name a few. I especially think that we can take some lessons and inspiration from the story of Erwin Schrödinger, whose audacious speculations and predictions in biology have been hugely influential. Physics has undergone exponential growth since Schrödinger's time, and in this "naive" spirit, I will playfully venture into ways in which biology may inform some mysteries in physics and vice versa. But first let's learn from Schrödinger's scientific audacity and ingenious contributions to biology.

Schrödinger moved to Ireland to escape the Nazis in 1938, and continued his work in Dublin. In 1943 he gave a series of lectures at Trinity College that would eventually revolutionize biology. In 1944 they were published in a tiny book called *What Is Life?* In it, Schrödinger speculated on how physics can synchronize with biology and chemistry to explain how life can emerge from inanimate matter. Schrödinger proposed his central question by asking, "How can the events in space and time which take place within the spatial boundary of a living organism be accounted for by physics and chemistry?"

In other words, how is it that the same laws of physics that describe a star account for the intricate processes of metabolism within the living cell? Schrödinger quickly admits that the physics of his time was insufficient to explain some of the ingenious experimental findings that his predecessors and contemporaries had made about the living cell. He even considered the possibility that physics as it was known then may not be enough, when he said, "One must be prepared to find a new physical law prevailing in it." Despite this, Schrödinger marches ahead, using the physics of his time to make some auspicious predictions that inspired the discovery of the double helix structure and functioning of DNA.

Quantum mechanics plays a key role in life, since it is the quantum that is necessary for the stability of atoms and the bonding rules to construct the plethora of molecules found in living matter. (As to whether the more "exotic" quantum properties, such as entanglement or quantum tunneling—where a wave function can actually pass through a barrier—may play a role in life, this remains an open research issue. We'll get to that.) Schrödinger opens his argument by conjuring quantum mechanics as the starting point to understand the difference between nonliving and living matter. For example, the bulk properties of a piece of metal, such as its rigidity and ability to conduct electrons, require an emergent long-range order, as we saw when I first raised the emergence principle. These properties should be a result of the bonding mechanism and the collective effects of the quantum wavelike properties of electrons in the metal's atoms. Schrödinger then describes how the atoms in inanimate matter can organize themselves spatially in a periodic crystal, before making a daring leap. Life clearly is more complicated and variable than a piece of metal, so periodicity isn't going to cut it. So Schrödinger makes a bold proposal: that some key processes in living matter should be governed by aperiodic crystals. More astonishing, Schrödinger postulates this nonrepetitive molecular structure—which will turn out to be a great description of DNA—should house a "code-script" that would give rise to "the entire pattern of the individual's future development and of its functioning in the mature state."

Before Schrödinger's time biologists had the idea of the gene, but it was a formless unit of inheritance, with much that was left unknown. When Schrödinger proposed his idea about how genetic material should work, it was completely unanticipated in anyone else's work. Today, of course, the idea that genes are governed by a code, similar to a computer code, which could program the structures and mechanisms of the cell and determine the fate of living organisms, might just seem to be common sense. While I say this, exactly how this is accomplished at a molecular level is still a rich research enterprise in biology. What is remarkable is that Schrödinger used reasoning stemming from quantum mechanics to formulate his hypothesis. Schrödinger was an outsider to biology, and this naturally made him a hidden deviant and ripe for making a paradigm-shifting contribution. Whether quantum mechanics was operating in some subtle way for life's processes was not central to Schrödinger's argument, but his new line of reasoning provided a completely new set of concepts to explore novel mechanisms. It's worth noting that no one took Schrödinger's "code-script" seriously, even after biologist Oswald Avery's 1944 publication of a paper that gave strong evidence for DNA as the carrier of genetic material. Of course, Avery's work itself wasn't immediately embraced: as described by science historian Matthew Cobb, part of the reason for this lack of excitement over Avery's discovery was that DNA was thought to be a "boring" molecule with a repetitive structure—exactly what Schrödinger had said a gene could not be. Nevertheless, Schrödinger's quantum reasoning led to a prediction that the aperiodic structure of DNA carried a code that could program life.

Schrödinger's unappreciated insight came about from an auspicious set of events. In 1943 the US Scientific Research and Development Committee hired some scientists to study information from radar for antiaircraft purposes. Among the scientists were Claude Shannon, the pioneer of information theory, and Norbert Wiener, who found connections between control systems in machines and biological life and coined his findings cybernetics. Both fields have surged into prominence today, especially in the form of quantum information theory, as well as machine learning and artificial intelligence, but they were hugely influential in our understanding of the theoretical and computational underpinnings of life.

As important as Shannon and Wiener are, the game really changed when computer science pioneer John von Neumann argued that a gene was the carrier of information. Von Neumann imagined a gene as a tape that could program an organism. He made an analogy between self-replicating machines and cellular replications. In this case, in order for a machine to replicate itself, there must be an underlying mechanism present to copy the information that specifies the machine itself. In a 1951 conference proceeding, the godfather of developmental biology Sydney Brenner said "[Von Neumann] divided the machine—the automaton as he called it—into three components: the functional part of the automaton; a decoding section which actually takes a tape, reads the instructions and builds the automaton; and a device that takes a copy of this tape and inserts it into the new automaton." It wasn't immediately obvious what von Neumann was onto, at least not to Brenner: "I think that because of the cultural differences between most biologists on the one hand, and physicists and mathematicians on the other, it had absolutely no impact at all. Of course, I wasn't smart enough to really see then that this is what DNA and the genetic code was all about."[1] Brenner's observations about how cultural differences between biologists, physicists, and mathematicians were in part responsible for biologists missing von Neumann's earth-shattering insight about information, genes, and replication remain relevant today. The silos of scientific disciplines and scientific social orders still limit how scientists work.

There were two more insights that Schrödinger posited between life and physics. I will leave it for you to read his landmark book

What Is Life? about his reasoning that living things have to negate the second law of thermodynamics, which describes the fact that closed systems evolve to maximize disorder, or entropy. Schrödinger called this negentropy: "What an organism feeds upon is negative entropy. Or, to put it less paradoxically, the essential thing in metabolism is that the organism succeeds in freeing itself from all the entropy it cannot help producing while alive." Living entities have to stabilize their structure and function over their lifetime against the fundamental tendency for disorder, since entropy always increases. Other biologists, mathematicians, and physicists have further developed Schrödinger's idea into rigorous mathematical statements about how this works.

His third insight has to do with consciousness, and we'll come back to that in the conclusion of the book. First I want to ask, how would Schrödinger revise *What Is Life?* with the new developments in modern biology and cosmology?

Over the years I have developed a pattern: I have chosen to make friends with biologists. Having those friendships means that I'm always asking about what they think is cool in biology. In the past several years I've had frequent conversations with my friend Salvador Almagro-Moreno, a molecular biologist. Sometimes meeting over a drink, we exchange ideas: I tell him my latest musings in cosmology research and he tells me about his musings in biology. It probably won't surprise you to learn that Salvador is a proud owner of a first edition of Schrödinger's *What Is Life?* He and I share a vision that I think many of our fellow biologists and physicists would find too deviant, and even repugnant. We don't let that stop us, and there have been many times that we stayed up late talking about this fascination. There was definitely some strategy here: in part our conversations were an exercise in deliberately generating an outsider perspective, hopefully to benefit each other's research insights. But it wasn't all so calculating: it was a lot of fun, and just one of the things that make our jobs as scientists so rewarding. What is interesting to me about our discussions is not just the ideas themselves, but how our conversations generated new questions in both our zones of inquiry.

In 2014, I held the E. E. Just chair in natural sciences at Dartmouth College. Ernest Everett Just was a pioneering Black developmental biologist who cemented the early field of epigenetics. His remarkable life story, scientific contributions, and legacy are majestically described in *Black Apollo of Science* by Kenneth Manning. As the E. E. Just professor, I had the responsibility of running a university-wide science program for promising students from underrepresented groups interested in science. Salvador was awarded the E. E. Just postdoctoral fellowship and helped me with the program by leading its mentoring and teaching component. Nowadays, Salvador is a biology professor at the University of Central Florida, one of the largest Hispanic-serving institutes in the United States, and is leading similar efforts as the E. E. Just program. One day we were discussing some program business when the conversation, as usual for us, evolved. I started telling my friend about a research idea I had in cosmology. The idea was a new mechanism to understand what is called the fine-tuning problem: as with the value of the cosmological constant, there are other constants of nature that have values that are just the way they need to be so that life could happen. A hugely important set of constants, known as coupling constants, that determine the strength of the force—such as electromagnetism or the weak force—are involved in interaction between particles. The idea I told Salvador about was that, if the universe went through a large number of cycles of collapse and expansion, the big bang phase would provide an opportunity to reset the values of the coupling constants. Most cycles of expansion might not host life, but eventually the universe would hit the jackpot. Salvador asked me very precise questions about my project, and the next day, in a state of elation, he told me that he was able to implement the cyclic-universe idea into the development of a theory based on some experiments he completed involving genetic evolution in biology.

One of Salvador's concerns was understanding what it took for bacteria to become pathogenic—capable of causing disease in a host like us. According to him, various combinations of genes, like a slot machine, can hit the jackpot in giving the bacteria the ability to become harmful. Salvador made a brilliant analogy between the coupling constants in my cosmological model and the various forms of a gene, known as alleles of the gene. He likened the bacterium's population cycles and environmental factors to a cyclic universe as a mechanism to change the different combination of genes. The different replicating populations represented the many cycles in the model, each with its own alleles or variations. This led him to propose the theory of virulence adaptive polymorphisms (VAPs).[2] The work was published in the top journal in microbiology. The idea worked and has made a big mark on his field.

Now it was time for my friend to return the favor and inspire some ideas for cosmology. Over the years we had mischievously developed a conviction and intuition that there is a hidden interdependence between living systems and cosmology. I came at the question from cosmology. Salvador came at it from a biological direction. And to our delight this question led us to a few issues that the big bang and living systems (such as ecosystems) have in common.

A major concern for both of us has to do with the flow of entropy in the universe, whether at biological or cosmological scales. In the epoch in the early universe before there were stars and planets, the universe was mostly filled with an equal amount of radiation and matter, where the photons and electrons were in thermal equilibrium. If a collection of gas molecules occupy a closed system, say a room, they are going to tend to thermal equilibrium—the temperature, essentially, will become the same throughout the room. As they approach equilibrium, their entropy will increase. Entropy is a measure of disorder, and so of ignorance about what the gas molecules are doing. The entropy will increase when there are more particles and more space for the particles to potentially occupy—it becomes increasingly difficult to specify what they are all doing. Entropy also negatively affects the ability of any system to do work. Physics, biology, and chemistry rely on an important concept called free energy, which is a measure of how much energy in a system (such as a living cell) is able to be used for work. Mathematically we can express free energy as $\Delta F = \Delta E - T\Delta S$. The equation states that a system can do work with a positive change in free energy (ΔF), where a positive contribution comes from a change in energy (ΔE) and a negative contribution from the change in entropy (ΔS). The entropy contribution to free energy reduces the amount of energy that can be used to do work. For example, sunlight shining on Earth generates free energy, which we calculate by adding the contribution of the potential energy stored in the wavelength of the photons and subtracting the entropy from the array of photons.

But there are some important caveats having to do with gravity. The gravitational expansion of the universe keeps the matter and photons in the cosmic microwave background homogeneously distributed. Situations of extremely high gravitational entropy are contained when matter collapses into localized objects like black holes, which did not exist in the era when the CMB formed. So, when the universe expands, gravity acts to distribute matter in a homogenous, ordered fashion, and this lowers the entropy of the universe. When gravity acts to coalesce matter into supermassive black holes, entropy goes up: the heavier the black holes the larger the entropy.[3]

And there is an important problem. The photons and matter in our universe were in equilibrium during the CMB epoch, so the entropy then was high. But as the universe continued to evolve, the entropy would continue to increase. This implies that the entropy of the very early universe, before the CMB epoch, must have been very low.

At the largest observable distances, we see a connected pattern of large-scale structure of galaxies distributed across the universe. As the universe continued to expand and cool, out-of-equilibrium structures with varying complexity like stars, clusters of galaxies, and life formed. The structures will contain lower entropy than the rest of the universe. By starting off with low entropy, the universe is able to arrest the growth of entropy against the trend of the second law, by concentrating regions of lower entropy within cosmic structures. These cosmic structures, such as stars, store potential energy; in the case of stars, it's from the rest mass of hydrogen, which will release highly energetic photons from nuclear fusion. This entropy-lowering network of structures becomes the main currency for the biosphere and life on planets. Even the father of thermodynamics, Ludwig Boltzmann, said, "The general struggle for existence of animate beings is therefore not a struggle for raw materials . . . nor for energy which exists in plenty in any body in the form of heat, but a struggle for entropy, which becomes available through the transition of energy from the hot sun to the cold earth."[4] Nevertheless, even as the universe deviated from homogeneity, by seeding and forming lower entropy structure, entropy elsewhere in the universe continued to grow. And entropy also has a tendency to grow within those structures. This makes entropy, or its absence, a key player in sustaining cosmic structure, such as stars and life; therefore, an early lifeless universe with low entropy is necessary for life here on Earth. Stars like our sun radiate free energy to the earth. This free energy is absorbed by electrons in plants and used for the necessary chemical work for its living function. The plant will release this energy in the form of heat and give off to the universe more entropy than it took in.

Unfortunately, it is difficult to explain with our current understanding of physics why the entropy was so low in the early universe. In fact, this problem of the low entropy we demand of the big bang is one of the major problems with the theory. It was first identified by Roger Penrose. Its solution remains a mystery.[5] I remember discussing this problem with Penrose on a nice summer walk, and we both agreed that if we wanted to understand how gravity could have helped set up this unlikely scenario, we were going to need the real connection between entropy and gravity, which is currently lacking, to reveal its nature. One hint is that at the earliest moments of the universe, close to the big bang, the curvature of space-time approached infinity. Whatever new physics tamed this infinity should tell us why the entropy of the universe was so low. We will get to this.

The biology side of the story stems from Salvador Almagro-Moreno's research into the genetic and ecological drivers that lead populations of harmless bacteria to evolve and emerge as pathogens.

FIGURE 25: A simulation of the large-scale structure of the universe. Each dot represents a galactic system.

Crucial to the story is that it isn't just a question of the genetic code of the bacteria. One of Salvador's mantras is that life is an adaptive phenomena responding to constant and unexpected changes in pressures from the environment. If life can have more channels and resources for being adaptative, it will find a way to use them. Central to his research is understanding evolution from the genetic code in a population of organisms, and the epigenetic influences from the ecosystem. Epigenetic factors are called that because they sit on top of genetic factors, and they are one of those other channels for adapting beyond changes to the genetic code. For example, an environmental factor, such as a pattern of electrical current hitting the cell membrane during replication, can enhance or suppress certain genetic factors, leading to completely different features in the phenotype of the offspring.

This makes an organism an emergent phenomenon, where the final shape of it is not contained in the individual pieces and influences that make it up. Recall that in emergent phenomenon in general, it is the interactions of the building blocks that collectively exhibit the emergence. This also implies that a population is emergent, too. Living things comprise a network of interactions that is mediated through the environment. A living system is able to regulate billions of cells to maintain its overall functioning. Beyond that, collections of organisms belong to a network called an ecosystem, which also maintains a dynamical equilibrium.

This extends all the way to networks at life's largest scales. The idea of the earth being a self-regulating ecosystem was codiscovered by James Lovelock and Lynn Margulis in the 1970s, and it became known as the Gaia theory. The name Gaia came from the goddess who personified Earth in Greek mythology. In response to the name, Lovelock, a chemist, observed that "biologists scorned it . . . it gradually became known as Earth Systems Science, but it is the same thing." Whatever you call it, the takeaway for me is that the flow of negative entropy exists not only for individual living things but for the entire earth. The sun sends free energy to the earth, and through a chain of complex interactions, the energy gets distributed through a network of interactions to living things, each relying on it to maintain its complexity in the face of increasing disorder. But there's no free lunch: when living things release this energy back into the environment, they mostly do so in a form that has higher entropy than what they received. Salvador and I noticed the uncanny parallels between living systems and the evolution of the universe through the lens of entropy.

This could seem like a coincidence in the behavior of the universe and of life, but we decided to treat the parallel as though it were not. Instead, we proposed that Schrödinger's idea of negentropy is one of the central organizing principles of the evolution of the cosmos and the existence of life. Salvador elected to call this the *entropocentric* principle, a wink at the anthropic principle that first emerged from string theory and caused such a controversy when I was first working on the vacuum-energy problem. The anthropic principle, in its strong form, states that the universe is fine-tuned for life. The laws of nature and values of coupling constants of their interactions have the values that are consistent with life on Earth. For example, if the strength of the nuclear interactions differed by a few percent then stars would not be able to produce carbon and there would be no carbon-based life. The fine-tuning problem may not be as severe as it seems. In research I conducted with my colleagues Fred Adams, Evan Grohs, and Laura Mersini-Houghton, we showed that the universe can be fit for life even when we let the constants of nature like gravity, vacuum, and electromagnetism vary, so long as they vary simultaneously.[6] Maybe we don't need the anthropic principle after all. The entropocentric principle, on the other hand, is harder to shake. If the universe was unable to provide pathways that enabled it to transfer regions of lower entropy, then life as we know it would not exist. We call this biological dependence on the entropic relationship of the cosmic structure the entropocentric universe. Living systems situate themselves to reduce their entropy by expelling it out into the environment, while consuming energy from their environment. Did the universe play the same game near the big bang?

13

DARK IDEAS ON ALIEN LIFE

Salvador Almagro-Moreno isn't the only person with whom I went exploring in new territory of the mind. When I lived in the Bay Area, I used to get together with my friend Jaron Lanier to explore the implications of spectacularly weird thought experiments. Occasionally, one of these conversations would lead to an interesting outcome. This chapter explores one of them. Outlandish thought experiments have been essential in the intellectual history of science, but the point isn't the weirdness itself. The payoff of thinking about strange things like Schrödinger's cat, the infamous cat that is alive and dead at the same time, is not necessarily that we should then "believe" in the existence of such a cat. Instead, we can hope that uncommon ideas will shed light on the murky margins of our thoughts; in the case of Schrödinger's cat, in dealing with the question of superposition. The point is not to confuse or bamboozle people, but to eventually find a way to think that makes more sense and is a little less murky.

The bizarre notion I want to consider here came from a discussion of the search for alien life forms. There are a variety of ways to look for signs of alien life in the universe, usually involving a large array of telescopes. One approach is founded on the hope that perhaps astronomers will get lucky and chance upon an alien radio broadcast. But in the thought experiment Lanier and I explored, we considered a different and far more dramatic possibility. Suppose that there are lots of alien civilizations running hugely capacious quantum computers of the sort that Google and others are just beginning to build here on Earth. This leads to a question of high weirdness: Would an extreme amount of very distant quantum computation result in any astronomically observable effect? Could we humans see evidence of a universe teeming with quantum computers by carefully examining the night sky?

We thought about various ways this might be possible, but in the end we focused on one wonderful possibility. So here it is: First, alien quantum computers could explain the mystery of dark energy, because computation by multitudes of alien creatures across the universe bends (or rather unbends) the universe as a whole. Because we can observe the effect of dark energy, accelerating the expansion of the universe, this implies that we have already seen evidence that our universe is alive beyond us—we just haven't recognized it as such! And we found, fortunately, that contemplating this almost imponderable notion has a human-scale practical payoff: it helps us clarify how we think about plausible relationships between gravity and quantum information. (If you think this is strange, you should read some of the competing ideas. One recent paper suggests that dark energy is actually a sign that time is about to cease to be time and turn into space instead. We'd then be frozen out of time, but be four dimensional. Compared to that, our proposal, aliens and all, is practically tame.)

Let's go through the argument step by step: What is a quantum computer and why would aliens be using them? Let's assume that, just like us, plenty of alien civilizations will want the best possible computers for some purpose or other. For the sake of argument, we'll assume the aliens want to enjoy high-quality virtual reality, and so they build computers to make that happen.

If the computers that run alien VR are of the classical kind we use these days (based on the mathematical framework laid down in the mid-twentieth century by computer science pioneers John von Neumann and Alan Turing), then aliens would generally endure an inferior sort of virtual-world experience. You might think that classical computers should be up to the task—after all, the special effects in movies are getting fairly realistic, and classical computers are able to calculate those effects—but they are not. Remember that movies are prepared in advance. Virtual reality, however, must create sensations for the human body on the fly and as quickly as reality does. Classical computers can't work that fast. Furthermore, there are cases where the human body is able to respond to reality at the highest possible level of sensitivity. For example, the retina can, in certain cases, generate a neural response to a single photon. In a case like that, the human body has become as discriminating as physics can possibly allow. Just as classical computers can't be as fast as the universe, they can't be that discerning, either. If we assume that aliens elsewhere also evolved to be as sensitive to this ultimate, quantum level of reality in some special cases as we are to light, then when they try to design a nonquantum supercomputer and VR apparatus that could simulate reality at the ultimate level of detail, they would have run into problems. That's one reason we guess that discerning aliens would seek the power of quantum computers to run their virtual worlds.

Quantum computers are not yet adequately developed for practical uses on Earth, but they have the theoretical potential to pulverize regular computers in a wide range of calculation contests. A quantum computer can work as if there were copies of it in many parallel universes at once, simultaneously exploring variations of the computational task at hand.

Suppose the computer is calculating what a virtual rose petal should feel like to your fingertip. The rose petal is pliant, so every part you touch changes all the other parts you touch. You have to calculate all the parts at once, and there's only a single solution that consistently reconciles all the tiny events in different locations of the petal so that it feels realistic. A quantum computer can be calculating a huge variety of different versions of the petal simultaneously, even though only one variation is the correct version. That correct version can then be instantly presented to your fingertip, perhaps by the "octopus butler robot" that Lanier has imagined in his book *Dawn of the New Everything*, as if the computer had somehow known which variation would be correct from the start.

One big engineering problem for quantum computer designers is heat. Heat is an almost universal problem for any computer designer. Every time you change a bit inside a computer, you're doing at least a little bit of work, whether that bit is implemented as a bead in an abacus or as a charge in a semiconductor in a silicon chip. Work always gives off heat.

Let's consider the example of the abacus. When you move a bead up and down you generate some heat from friction on the wire the bead is sliding on. If you do this only a few times a second, you won't even notice that heat, but if you move the bead millions of times a second you will melt the wire. Now consider a quantum abacus. This would be a little like having a bunch of copies of the abacus in different parallel realities, each with the beads in different positions, each exploring a different variation of a problem. You can think of each individual bead as being like Schrödinger's cat: you can either think of a bead as being both up and down at the same time, or that in each particular universe it is either up *or* down. We call this kind of bead a qbit (quantum bit) instead of a bit. If the quantum abacus gets hot, the beads start jiggling, in just the way that oil in a hot frying pan will start sizzling. If the beads jiggle too much, it becomes impossible to say which universe has a bead that is up or down, which means that the differences between the states of the beads in the parallel universes disappear. When that happens, the quantum advantage also disappears.

The role of the beads in a quantum abacus is usually performed by fundamental particles, which means that even a tiny amount of heat will have this ruinous effect. Quantum computer designers here on Earth have been forced to advance the state of the art of extreme refrigeration. The machines have to be run at ultracold temperatures.

There are other ways to deal with the issue, however. One potential way to reduce the heat problem is with a topological design. Topology is a field of mathematics that describes how things connect. For instance, if you put a bead on a circular wire, the rules of the game change because pushing up on a bead that is already up will force it around to become down. The meaning of an "up push" becomes dependent on the previous push. When you change the way things connect, you change the kinds of information that can be contained in those things. An example from human cultural history is that the ancient Incas used elaborate knots called *khipu* as a record-keeping scheme and a form of written language.

If you use topology to hold information, you don't have to worry about heat quite as much as you do when you are moving bits around within a fixed topology. Instead of just moving a bead, for example, you could also bend and change the connections of the wire so that the bead moves in different ways. In addition to loops, you might explore various knots, branching structures, and so on. Enough heat could still melt the wire and ruin your topology, but if the heat is only enough to jiggle the beads, then the topology won't be harmed and you can preserve a lot of information.

Physicists like Nobel laureate Frank Wilczek are trying to figure out how to make a practical topological quantum computer on Earth. The general idea is to move around artificial, flat, fundamental particles called anyons (which we first encountered in Jim Gates's work) within tiny confines so as to tie knots with the paths that they trace as they move. This is a remarkable idea because the knots are only knots if you think of time as one of the dimensions within which the strands are held tight.

So far, so good: none of these ideas about quantum computers is anything other than mainstream, not radical. But Lanier and I went a step further to propose that alien computers are not only topological quantum machines, but that they are gravitational. The particles they tie knots with are gravitons. And this takes quantum computers into the realm of dark energy.

As we have discussed, dark energy is treated as synonymous with the idea of the cosmological constant these days. That wasn't always the case. When Einstein formulated the theory of general relativity in 1915, the math predicted that the universe should be expanding. That seemed wrong, so Einstein added the cosmological constant so the universe would be static. Then in 1925 Edwin Hubble gathered data that showed that the universe actually was expanding, so Einstein got rid of the constant. Then, in 1998, our friend Berkeley astronomer Saul Perlmutter gathered even better data and showed that not only is the universe expanding, but the rate of expansion is accelerating. And so physicists brought the cosmological constant back, but with a new value. And that's what we call dark energy, and in case you forgot, it's weird: it behaves like a repulsive fluid with negative pressure that fills all the space around us, pulling everything apart. Given a choice, it doesn't sound like the sort of fluid you'd want to swim in, but we are swimming in it.

As we have seen, at present there is no solution to the dark energy problem. Prominent physicists like Ed Witten have called it the greatest embarrassment of theoretical physics. It has forced physicists back to the drawing board—all the way back to the foundations of quantum mechanics and general relativity. And for me and Jaron Lanier, that meant quantum computers.

Before we get to the alien computers, however, I want to consider one last difference between Einstein's general relativity and quantum field theory that has kept physicists from reconciling them. That is the passage of time. In general relativity there is no master clock, but in quantum field theory time is expressed using a master clock that sort of lives off to the side in a metaphysical realm. There are certain phenomena, like black holes and the vacuum, which can be approached by either theory. In the case of the vacuum, we can clearly see some ways in which quantum field theory is unsatisfying. For instance, remember that general relativity is all about curved space-time. In fact, the cosmological constant is just an adjustment of the curvature. It turns out that a detailed mathematical analysis reveals that quantum fields in the curved spaces of general relativity don't have unique quantum states. An example of this strange state of affairs is called the Unruh effect, named after Bill Unruh, a physicist who described it, which predicts that the vacuum will appear hotter to an accelerating observer. If the effect is eventually observed, then most of what we write here will be rendered wrong, but for the purposes of our argument we're assuming the prediction is wrong and that the nonrelativistic inner self of quantum field theory is at the root of the mistake.

Recall one of the usual images invoked to describe general relativity: if you try to use a passing train naively as a pitch pipe, you'll inevitably sing out of tune. This is because the train's whistle seems to go up and down in pitch as it flies by. What's really going on is that sound waves wash over you in different concentrations depending on where they were emitted as the train travels relative to you. When they get bunched up, because the train is approaching, they seem to be higher in pitch. That creates an illusion of a changing tone when the train goes by.

In the same way, quantum field theory asks a traveler in an accelerating spaceship to consider an absolute clock, but that clock's time cannot be taken at face value, just like the sound of the moving train whistle can't be used as a steady reference. Screwing up the measurement of time screws up everything else. The vacuum should seem hot, according to quantum field theory, for the same reason the whistle seems to get higher. Relativity doesn't make this mistake, but quantum mechanics, the grandfather of quantum field theory, has passed on this "bad gene." Within the context of quantum field theory, we don't have a truly relativistic platform with which to describe the vacuum.

Back to the possibility of gravitonic topological alien computers! Recall the earlier image of the many copies of a quantum abacus, distributed across many realities, all working in parallel. When we move to topological abacuses, where the wires can be tied into loops, trees, knots, and so on, what is going on in all those parallel universes? There is an interesting divergence. On one hand, there can be a bead that is jiggling between the up and down positions on a wire. But although topological quantum computers would rely on changing the shapes of the wires, the shapes themselves are definite. They cannot jiggle themselves into loops that are simultaneously open and closed, in what you might call "sort of loops." A wire is either a loop or it isn't.

The interesting thing to notice is what happens if you superimpose all the versions of a quantum abacus from the various realities it's in. The superposition changes depending on the role that topology plays in your computer. The complexity of a nontopological (nonknotted) superposition is linear, which means simply that it's exactly as complicated as you'd expect from summing up the complexity of its parts. If, however, you superimpose knotted, topologically interesting abacuses, the result will not be linear, because not all the intermediate states are possible. If you superimpose a loop and a nonloop, you don't get a distribution of possibilities including a "sort of loop."

Now recall that we just mentioned something else that isn't linear: relativity. Space-time is curved, not linear. This is what we observed when criticizing the prediction of the Unruh effect. The quantum state of gravity, which can be thought of as the superposition of the quantum abacuses that either are or describe (depending on your philosophical preference) the state of space-time and/or gravity in the universe, is not linear. Could that be because gravity's abacuses are knotted?

So here's a scenario of what might have happened in the history of our universe. These days we take matter for granted as a dominant component of reality, but when the universe was younger, it was too hot for matter to be commonplace. Eventually the universe cooled down and matter became an important phenomenon. With matter came chemistry, and with chemistry came life and evolution, leading to smart aliens who experienced their own Moore's laws, which drove them to develop gravitonic quantum computers.

As it happens, the amount of dark energy in the universe reduced rapidly just as matter became important. In terms of our thought experiment, that was not a coincidence. It was because the aliens used dark energy as a resource to run their ultimate computers in much the same way we devour oil to run our cars and jets.

Before we continue, there is an even more daunting issue that sneaks up on anyone who tries to explain away the cosmological constant problem: Why now? I had tried to tackle this when I was at Stanford. The surprising thing is that today the cosmological constant/dark energy is not zero but proportional to the dense matter; it seems to be tracking the amount of matter. The approach Lanier and I took in our thought experiment was to turn the question of "Why now?" into a new form: Why here? Recently a consistent cosmology with no dark energy has been proposed by a large number of respected cosmologists, like Joe Silk and Subir Sarkar, who argue that if we live in a region void of the excess dark matter then we can do away with dark energy. Likewise, our aliens come from a region void of dark energy because this void represents a biosphere of computational activity. According to a detailed multidata analysis of the latest measurement of the cosmic microwave background, the Sloan Digital Sky Survey and type Ia supernovae observations, such a region exists. Working with my colleagues at Penn State and CERN, we were able to show that all that data is consistent with a region void of dark energy spanning some two hundred megaparsec. This is a huge region; for comparison, ten kiloparsec is about the radius of a typical spiral galaxy like our own Milky Way, and a parsec itself is already pretty big, at more than three light-years. So we at least have a spot where these aliens could live with their computers. Everywhere outside the aliens' existence, where there are no computers, the dark energy would exist.

How would this work? In our idea, aliens use the vacuum state as a "reservoir" of qbits. As they do computation, they tie more and more complicated knots in the gravitational, or space-time, quantum state of the universe, which we can think of as all those knotted, superimposed abacuses. This has the effect of "using up" the curvature due to the vacuum energy.

Thus, instead of seeing the enormous value we would predict naively from thinking about a vacuum state that hadn't been tampered with, we instead see the tiny cosmological constant that Saul Perlmutter measured. If this is correct, then that would mean that the aliens have almost, but not quite, maxed out the computational potential of the universe!

Our thought experiment didn't quite lead us to an idea of how alien gravitonic computers could work, but we do have some hints. Vacuum energy universally interacts with the gravitational field. There are some important direct channels through which this interaction between gravity and the vacuum energy takes place. It was shown long ago by Einstein and Élie Cartan that fermionic matter (like electrons) will universally mediate vacuum energy with gravity. But there are many types of fermions, of course, and theorists have shown that chargeless fermions (similar to neutrinos) will have the most resonance with the vacuum energy. To highlight the computational mechanism, consider an analogy with superfluid helium. Helium atoms behave like inert fermions, and at very low temperatures a quantum interaction between the helium atoms make them condense into a superfluid with astonishing emergent properties; one is that the superfluid has a negative pressure, just like we considered before for vacuum energy. A quantum field theory calculation successfully predicts this about helium, and Nobel Prizes were earned for the experimental verification of this superfluid state. It turns out that coincidentally gravity mediates exactly the same type of quantum interaction between chargeless fermions!

Therefore the analogy between superfluid helium is quite relevant and important for us because, if a region of space-time is cold enough, the correlations between inert fermionic matter can act as a self-organized medium that tethers the ribbonlike structure of the gravitational-fermionic vacuum energy.[1] Very intelligent aliens can achieve quantum computation by exciting the various energy and spin states in this fermion-gravitonic superfluid. How cold does this fluid environment have to be? A back-of-the-envelope calculation reveals that qbit states can be manipulated by turning vacuum energy into thermal radiation so long as the temperature of the superfluid is less than the temperature of the cosmic microwave background, three degrees above absolute zero. However, the region will have to be filled with the neutral fermionic matter that the aliens use to fuel the process. This material can be gotten from supernovae, which emit a huge flux of neutrinos suitable for the purpose. Neutrinos are famously hard to catch, so the aliens will either have to figure out how to trap and contain the neutrinos that pass through their part of the universe, or simply have to inhabit regions rich with supernovae.

It is plausible that the aliens would have control of hot rod gravitational wave detectors, not unlike the laser-interferometer gravitational-wave observatory, or LIGO, that detected gravitational waves first in 2015, that can detect and write information into the topological vacuum state using coherent gravitons, in the same way that our transistors can detect electron currents and switch gates.

The gate switching would be powered by the gravitons' states emitted and absorbed from the vacuum. There have been other proposals to read and write gravity waves. UC Merced physicist Raymond Chiao has proposed a variety of curiously shaped superconducting antennas that might be able to read and write gravity waves.

To push weirdness to the extreme: maybe there could eventually be some way to safely create and manipulate artificial black holes that are analogous to anyons. Maybe moving them around carefully could perform quantum computation. At any rate, ideas about the actual implementation of a topological graviton computer are entirely speculative.

One oddity of gravitonic computers is that while the aliens and the interfaces they use to operate their computers exist locally, just like ordinary objects or like our bodies, the computation itself is not localized. The vacuum state is necessarily nonlocal since the qbits are entangled across our cosmic horizon (the portion of the universe that can be relevant to us as limited by the speed of light), and that region spans three thousand megaparsecs. That's one big machine room!

It should be, because the computers are stupendous. Paola A. Zizzi of the Università di Padova has calculated that universal information capacity in vacuum energy is approximately 10,120 bits. That's a lot of bits. How many qbits of computation are required by alien virtual worlds? If we guess how many civilizations there are, we can then estimate the average size of a planetary gravitonic computer. If we conservatively assume just one gravitonic computer per galaxy, we end up with googol-scale capacity—10,100 qbits—for an average alien gravitonic computer. When we try to fantasize what seems to us to be the absolutely ultimate planetary VR computer, assuming that Moore's law will run into the twenty-second century, we still only come up with a need for a capacity that is perhaps sixty orders of magnitude smaller.

But it's also true that you can never have enough computation! In fact, there's a reason alien computers might need to be so huge, and it has to do with heat. Gravitonic computers give off heat just like other computers. Where does that heat go? It is interesting that there are a vast number of highly energetic cosmic ray events and their source remains a mystery. Even so, there isn't anywhere near as much heat as might be expected from computers of the stupendous size hypothesized here.

You might just take this as a sign that the computers don't exist, but let's keep working with the idea that they do. If you want to reduce the heat a computer generates and you have a huge amount of memory, you have an amazing design option, which is called a reversible computer. That means that you change each bit in the whole computer only once, and then move on to another bit. That results in a total record of all computation—and that's why it's possible to run the computer in reverse: nothing has been lost. If you move each bead in an extremely capacious abacus only once, you don't generate the heat you would by moving each bead repeatedly. You can think of it as saving all the information in a tidy way instead of dispersing it. This is also a nice example of Claude Shannon's famous principle that information and entropy are related.

What will become of our cosmic khipu weavers? They might end up like Maxwell's demon, a thought experiment important to the development of the second law of thermodynamics that remains important precisely because Maxwell's demon can never exist. Or maybe the weavers do exist and we'll meet them. Maybe we'll eventually weave our own gravitonic khipu.

Of course it's possible the aliens don't exist but that graviton weaving will still turn out to be part of the solution to the mystery of dark energy. But that would suggest a natural, self-propelling process that relies on information storage—and that sounds like a definition of something very much like life.

14

INTO THE COSMIC MATRIX

All living things are born and change throughout their lifetime. The expanding universe is like this. Because the universe's space is expanding in time, when we reverse the clock, it returns to an epoch where time itself was born. For many years, I tried to get my mind around the question: What can exist if time ceases to exist?

To delve deeper into this and other issues about the early universe, it's useful to address a few common misconceptions about the big bang itself. Perhaps the biggest misunderstanding is that the big bang was some cataclysmic explosion that fueled the universe's expansion, where galaxies went flinging away from each other. In the standard big bang theory predicted by general relativity, the universe is assumed to be filled with a hot and dense gas of matter and radiation, whose energy and pressure source the expansion of spacetime itself. We've discussed before that general relativity provides an expanding solution to its equations; called the FRW model, for its discoverers Alexander Friedmann, Howard Robertson, and Arthur Walker, it was for a long time the standard model of cosmology. According to that solution to the equations of general relativity, the expansion of space is not an "explosion" but actually gradual in time.

Even more interesting is that time ends in a singularity of infinite curvature and energy density. What is this singularity really trying to tell us?

To try to answer the above question, Stephen Hawking and Roger Penrose provided a theorem that the FRW expanding solution will always suffer a singularity. At the singularity, the curvature goes to infinity, marking a breakdown of general relativity as a valid description of the space-time structure. The Hawking-Penrose theorem was based on a powerful equation discovered simultaneously by Indian theorist Amal Kumar Raychaudhuri and Soviet theorist Lev Landau. These Landau-Raychaudhuri equations relate paths of observers in a curved space-time, called geodesics, to singularities. Hawking and Penrose implemented these equations to prove that geodesics in an expanding space-time will exhibit pathologies as they approach the infinite curvature singularity at the beginning of the big bang. A useful set of geodesics to use as a diagnostic for singularities are light rays. In a flat space-time two light rays will follow parallel lines and never cross each other. In a curved space-time the paths of light rays can twist, focus, or even diverge, depending on the warping of the space-time. Hawking and Penrose showed that as light rays approach the earliest times toward the infinite past in an expanding space-time, their geodesics terminate. This geodesic incompleteness signals the infinite curvature and a singularity as time goes to zero.

A similar story happens for black hole singularities with geodesics, and many interpret singularities as a breakdown in the validity of general relativity at the singularity. Of course, all theorems, including the Hawking-Penrose theorem, do come with assumptions, or axioms, and axioms aren't necessarily true. Sometimes they can be relaxed; you may know that the parallel postulate in Euclidean geometry can be relaxed, and rather than breaking geometry, actually points the way to two other kinds, hyperbolic and elliptic. Perhaps if we can relax one of them this could be a way out of the inevitability of the cosmic singularity. One assumption of the Hawking-Penrose theorem is that matter is classical. Keep in mind, though, that the classical matter in the world is made up from quantum matter. And cosmic inflation uses quantum matter and the inflation field, and it may help resolve the singularity. But even the quantum state of inflation does not rescue the universe from the singularity.[1] The onset of inflation still remains unresolved. Over the last four decades cosmologists have been investigating various mechanisms based on new physics that could alleviate the big bang singularity. There are a handful of promising approaches my colleagues and I have pursued.

Aside from potentially satisfying the urge to understand the birth of the universe, pursuing the problems facing big bang physics provides an opportunity for physicists to guide themselves toward a theory of quantum gravity and to test the validity of those they find. At one level we have constructed theories of quantum gravity by attempting to make the principles of general relativity and quantum mechanics consistent with each other. But the issues that the early universe presents can serve as guideposts to a more fundamental theory—and it may not even be quantum gravity. As we go back in time, the universe approaches a length scale called the Planck length, which is thirty-five orders of magnitude smaller than a meter. Here, we expect gravity and quantum mechanics to speak to each other. In some quantum theories of gravity, space-time emerges from a pregeometric phase, a vision of reality that would force us to conceptualize matters where there is no space-time to refer to. In other models, the expanding universe must arise from primordial atoms of space-time. In those cases how these "atoms" interact to give us an expanding universe will be the prize hunt. As we explored, both string theory and loop quantum gravity have ingredients to avoid certain space-time singularities, and models of the early universe have been proposed and developed to transcend the singularity.

Understanding the evolution of the earliest universe provides the foundation for a sequence of important events for other forms of evolution to take place, such as star formation, which is central to life as we know it. Therefore, understanding the earliest stages and ideally the origin of the universe may give us new insights into the universe we currently inhabit. Just like the genome reveals new secrets about an organism, could it be that the pre–big bang universe can shed light on our universe in its current epoch?

Recall that a successful theory of the early universe must solve a handful of cosmological problems that currently have no confirmed solution. A suite of observations have made it clear what these conundrums are. In addition to the big bang singularity problem, these are the horizon and flatness problems (which João Magueijo, among others we've met, are working on), the fine-tuning problem, as well as the question of the origin of large-scale structure such as galaxies and the CMB. We have seen how cosmic inflation, without departing too much from the principles of general relativity and quantum field theory, has already been able to solve some of these problems, providing answers to the horizon and flatness problems, as well as giving an explanation for the universe's large-scale structure. You may wonder, if cosmic inflation does such a great job of explaining our current universe, then why seek alternatives? One answer is that it can't explain the singularity and fine-tuning problems. Another is that, even if an alternative theory turns out to be wrong or simply less successful than the explanations we already have, exploring it can still give us a new perspective that can improve the more conventional theory. What's more, I believe that the singularity problem that plagues inflation points to the source of all the cosmological problems—if we can resolve singularities, we might resolve it all.

Over the years my colleagues and I tried to alleviate some conceptual and mathematical problems that plagued cosmic inflation by constructing a superstring theory inflationary model, and we kept hitting roadblocks. It was only when we researched alternatives to inflation within the context of superstring theory that we got new insights into moving forward with the inflationary roadblocks—an example of the value of gaining an outsider's perspective. The fact that the universe is expanding actually limits the options as to what may have occurred before it started to expand. This boils down to the question of whether the universe had a beginning or is eternal.

We will now venture into discussing promising theories for understanding how the earliest stages of the universe could have emerged. Keep in mind that the theories that we will consider may or may not need inflation. And as we venture to get a sense of the future of the physics of the early universe, we should take stock of the two general and divergent hypotheses behind approaching a theory of quantum gravity.

Hypothesis 1: We should quantize gravity the same way that proved to be successful in other systems, such as quantum electrodynamics and quantum chromodynamics. In this approach, one identifies the classical system, such as the state space (or phase space) and applies rules for quantization. For example, in particle mechanics the phase space will comprise all positions and momenta of the dynamical particles. These measures commute with each other, which means that the order in which we multiply with them doesn't change the outcome.[2] Quantization rules impose that position and momenta don't commute, however, and the classical phase space gets promoted to a Hilbert space. In Hilbert space, the central objects are not finite vectors (of the kind that described momentum in our discussion of phase space early in the book), but rather infinite dimensional vectors. These are otherwise known as wave functions and correspond to a probability distribution.

Hypothesis 2: General relativity is a classical theory that should not be directly quantized according to hypothesis 1. Instead, at shorter distance scales there are more fundamental quantum degrees of freedom that give general relativity as a long-distance, low-energy classical theory: this is the quantum emergent principle at work. This is similar to how fluid dynamics emerges as a long-range theory of interacting atoms. As we discussed, string theory does not quantize general relativity, but general relativity instead emerges as a low-energy effective theory (that is, a theory that describes effects but not causes of an observed phenomenon—in this case, gravity), albeit in ten dimensions.

We've explored both so far in this book. But let's turn now to a promising approach to the early universe that pursues new directions. One of the interesting things about this approach is that it addresses primarily the big bang singularity and the other cosmological problems that cosmic inflation fails to address. This isn't unique to cosmic inflation: We will see that all approaches are limited and require concepts and tools that go beyond the current theories that we are considering.

In a pioneering publication in 1989, two theoretical physicists, Robert Brandenberger and Cumrun Vafa, presented a new approach to the early universe by attempting to solve three questions in one fell swoop. The key insight into the BV mechanism, as their result came to be called, harkens back to the concept of duality in quantum mechanics. They not only questioned the initial spatial singularity and infinite temperature at the bang but also related the solutions to those problems with something we take for granted: the dimensionality of space. The basic laws as we know them would be different if space were not three-dimensional. Electric and gravitational fields would not fall off inversely proportional to the square of the distance, which would affect all the chemistry necessary for the world as we know it. If we lived in two spatial dimensions, for example, life would not exist as we know it; for the function of carbon-based life depends on three-dimensional structure of folded proteins. In the spirit of relativity, we can think that the dimensionality of spacetime is not absolute and treat the fact that we live in three dimensions as a physical condition that emerged in the early universe.

Of course we've seen that string theory is a quantum theory of gravity that requires nine spatial dimensions in order to be quantum mechanically consistent. What Brandenberger and Vafa realized was that string theory had the correct ingredients to address the other cosmological problems in one fell swoop by first asking how the unique properties and symmetries of quantum strings address these cosmological conundrums. Consistent with superstring theory—a theory that fuses supersymmetry with string theory—Brandenberger and Vafa considered an early universe in nine spatial dimensions. Instead of the early universe being occupied with a thermal state of particles and radiation, it is filled with a thermal state of strings. Because strings are extended objects, and unlike particles, they have the special property that they can wind around a spatial direction. So, string theory has both oscillatory strings, and wound-up strings called winding strings. To get a feel for the differences between those two types of strings versus particles, consider the geometry of a two-dimensional torus, which looks like a doughnut. A particle can only move along the surface of the doughnut, but a string can both move along the surface of the torus and wrap around one of the cycles of the torus. These winding states, like rubber bands, have tension energy; if they are left to their devices they will cause the torus to collapse. This is also true if you are speaking about the kind of hypertorus you'd find in a nine-dimensional world such as Brandenberger and Vafa were considering. But according to general relativity the radiation that lives in space makes space expand. So, in a stringy universe there exist three types of matter: winding strings, string loops, and radiation. In such a universe space will expand and the winding energy will force the space to collapse after it expands too much.

Under ordinary circumstances, according to general relativity, such a universe will not be able to avoid a singularity. But string theory has a new ingredient that tells a different story: a symmetry we previously discussed (I used it in my thesis) called target space duality, or T-duality. This symmetry states that physics in the smallest possible region of space is equivalent to physics in the largest possible region of space. This happens because string theory has both matter and winding configurations present. If we perform a transformation that swaps the roles of these two states, it is equivalent to swapping motion in a large region of space and a small region of space. For example, if the torus universe has a radius R then the transformation will take R to 1/R; nevertheless, string theory remains unchanged. If the radius is cosmologically large, then strings moving will experience the same physics as if they were in a small space 1/R.

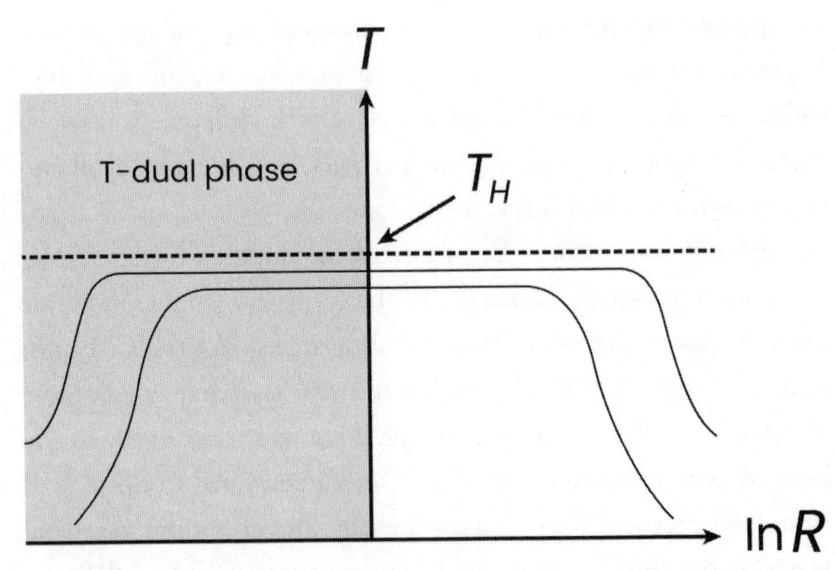

FIGURE 26: This diagram shows how the BV mechanism avoids the singularity as the radius goes to zero. We enter the T-dual phase instead of diverging with infinite curvature.

Here is an intuitive way to understand this. Let's imagine that the universe shrinks to distance scales that are microscopically small, such as the string length. In the standard big bang picture, because only particles were present, the temperature would tend to infinity as the universe got smaller—exactly the kind of singularity we'd like to avoid. But in a stringy universe as the temperature gets high enough as the scale shrinks, the energy of particles starts getting fed into the string oscillations. Eventually all these oscillation states get occupied and the universe approaches not an infinite temperature but a maximum possible temperature, known as the Hagedorn temperature. At the same time as we approached the R=0 singularity, there is a dual description of the universe in which the physics is the same as having winding modes moving in a large nonsingular universe. In a nutshell, the Brandenberger-Vafa mechanism prevents the universe from reaching the singularity by having a symmetry in the physics between winding string modes and particle modes and swapping small with large distances. If the universe can never become infinitely large, then by symmetry it can never go to zero length and will avoid the spatial singularity found in standard big bang cosmology.

But by using the same physics of winding states, the BV mechanism goes beyond this by predicting why we live in three spatial dimensions. Again, the argument uses the properties special to strings. Understanding why we live in three dimensions is as simple as understanding the question: In what dimension are two-point particles most likely to collide? In three dimensions if we set two particles (particles have zero dimension; strings have one dimension) off in a random direction, they are less likely to run into each other than in two dimensions. But in one dimension then the particles are guaranteed to run into each other.[3] A similar argument applies to strings, which are one-dimensional objects. It turns out that in three dimensions strings are most likely to collide with each other without the collisions being unavoidable; in anything larger than three spatial dimensions, strings are likely to never meet. In the Brandenberger-Vafa model there are also strings that wind in opposite directions relative to each other. If these strings run into each other they, like particles and antiparticles, will annihilate into radiation. Consider a three-dimensional torus of both winding strings, which keep the space from expanding due to their tension exerted on the space. If these winding modes annihilate then the space will expand according to general relativity. In a ten-dimensional universe, Brandenberger and Vafa found that the strings in the extra six dimensions never get to annihilate. As a result, three dimensions become our observable expanding universe while the other six remain wound up by the winding modes.

So now the BV mechanism tells us that our universe is a ten-dimensional, string-dominated space-time with all dimensions starting out microscopic. Applying the principle of maximal symmetry, at the beginning all space-time dimensions and string states are on equal footing.[4] Therefore, strings and antistrings will wind around all nine spatial dimensions. These dimensions will remain tiny until winding strings annihilate each other in three spatial dimensions. This will cause a large three-dimensional space to expand. Because of T-duality and the finite Hagedorn temperature, there is no big bang singularity. This suggests the universe did not emerge from a

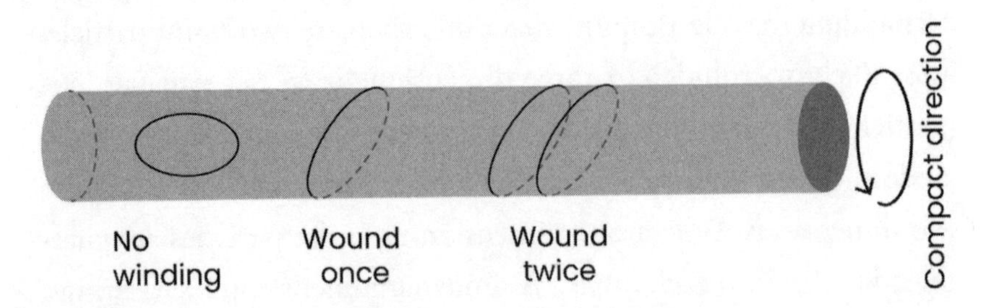

FIGURE 27: A schematic representation of a dimension being compactified by winding strings.

big bang singularity (that is, from a situation where the radius was zero and then began to increase).

The BV mechanism also provides a way of seeding the large-scale structure in the universe that is different from the way that inflation describes the process. In inflation cosmology it is quantum vacuum fluctuations that initiate cosmic structure. In the BV mechanism it is the thermal waves generated by the undulating gas of strings that give the observed nearly scale-invariant fluctuations seen in the CMB. Both mechanisms generate a spectrum of gravitational ripples of space-time that leave an imprint on the polarization of light in the CMB. These gravitational waves are predicted to leave in their wake a pattern on the CMB photons by creating an overall curling, like a pinwheel, of the light's polarization, called B-mode. Observational cosmologists have been on a hunt to find this faint signal, pushing the envelope of the most advanced detection technology known. Currently the Simons Array telescope led by my colleague and friend Brian Keating is being built to finally detect the primordial gravitational waves. But although both models predict the waves, they don't predict the same pattern in the B-mode polarization; observing the pattern could tell us if our universe underwent inflation or something more akin to BV. The predictions of gravitational waves between inflation and BV differ in that the shape of the power spectrum tilts in the opposite direction. In inflation the power spectrum is said to be red. This simply means that the longer wavelength perturbations have slightly less power than the shorter wavelength ones. BV has an opposite blue power spectrum.

The BV mechanism elegantly paves a way to solving another big problem in string theory that I discussed in a previous chapter on quantum gravity. Recall that the information about the extra dimensions in string theory show up in our four-dimensional world as a large number of fields that could have disastrous effects in our world, called moduli. If string theory is correct, it must provide a way to freeze, or stabilize, these moduli so that they do not disrupt the observations of the CMB and large-scale structure formation.

My colleague Subodh Patil and his collaborators elegantly discovered an intrinsically stringy mechanism such that the string modes generate a force to stabilize all the moduli in string theory, which made use of new symmetries that are unique to string theory. Patil is a bassist, and I am curious if he imagined the moduli stabilization mechanism as analogous to how a bass line can serve as an anchor for a melodic line. The mechanism also addresses a worry that I always had about the BV mechanism, one that I remained silent about for years. Superstring theory relies on supersymmetry; a symmetry puts fermions and bosons on the same footing, for its consistency. However, our world is not supersymmetric since matter is distinct from its force carriers. In particular, a time-dependent space-time such as our expanding universe is not consistent with supersymmetry. So, trusting the validity of these intrinsically supersymmetric string modes within a cosmological background seems a bit ad hoc. However, Patil and collaborators showed that it was exactly the breaking of supersymmetry that provided the forces for the stabilization of moduli.

The BV mechanism provides a compelling framework that successfully goes beyond the standard big bang theory. Some unresolved issues persist because the BV model assumes the string universe will exist in the Hagedorn phase for an indeterminate amount of time, then at some point in time the winding modes annihilate, and the universe begins to expand from a nonsingular state. How long does the universe remain in the high-energy stringy state? What placed the pre-expansion universe in such a state? The question of initial conditions persists in this version of the BV mechanism. One possible way out is to invoke a cyclic cosmology.[5] In this framework the universe undergoes a series of expansions and contractions long enough for the winding modes in three spatial dimensions to annihilate, and our three-dimensional universe expands and gets macroscopically larger than the other six.

Current technical and conceptual details are being researched that also touch on other profound questions about the early universe. One big issue that goes beyond BV and other approaches to the early universe concerns the emergence of space itself. In fact this subject is currently the state of the art in quantum gravity research. There are a handful of promising approaches to emergent space-time but I will restrict our discussion to an avenue that extends the BV approach naturally, while also illustrating how to improve a theoretical model. There currently exists no complete theory from which we can obtain an expanding cosmology from a pregeometric state, so I will argue that a promising direction comes from the general framework of noncommutative geometry. But what would we learn about cosmology from a space-time that emerges from such a pregeometric phase?

In 1989 my friend the late renowned theoretical physicist Joe Polchinski realized that strings were not the only fundamental objects in string theory. He did this by solving a long-standing problem about how to apply T-duality to a version of strings that have no windings. These are called open strings. Polchinski discovered new extended objects that are membrane-like, which he crowned Dirichlet-branes (or D-branes), that open strings can collectively end on.[6] These D-branes come in many dimensions; for example, a 2-brane is a two-dimensional hypersurface otherwise called a membrane. For part of my PhD dissertation work, I, along with Damien Easson and Robert Brandenberger, showed that the BV mechanism works even when we include a spectrum of D-brane states in string theory.[7] When I gave my thesis defense, one of my examiners was

David Lowe. Lowe is a string theorist, and he asked why we did not include D0-branes. This is a zero-dimensional object that strings attach to. I did not have a good answer, except that D0-branes do not have winding modes, like their higher-dimensional cousins. But I've reconsidered my response. It turns out that going down that path, incorporating D0-branes to reimagine the early universe, reveals an interesting new possibility—that the universe could have existed without reference to space itself.

The theory that describes D0-branes falls into a class of quantum geometric theories called noncommutative geometry. What is intriguing is that a handful of approaches to quantum gravity all have some semblance to a pre-space, where geometry itself is fuzzy, or noncommutative.

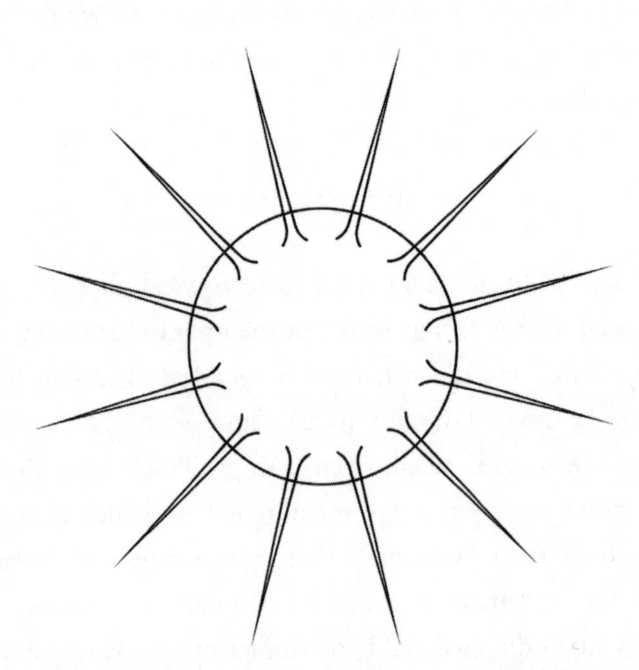

FIGURE 28: A membrane is classically unstable because it is more energetically favorable for infinite spikes to form on the surface than to keep the membrane surface smooth. This instability posed problems in quantizing the membrane as a fundamental object in quantum gravity.

In ordinary quantum mechanics the spin of the electron in the x and y direction cannot be known simultaneously. For example, the more certain you are about spin x the less certain you are about spin y. Suppose the determination of different directions in space were noncommuting. Then the usual notion of a smooth, continuous space-time is no longer valid. Imagine a noncommutative space where a location is restricted to be at a unique point in the x direction, then a definite location in the y direction will be uncertain. In general, a location in space becomes fuzzy in a noncommutative space. There are a handful of proposals in formulating noncommutative space-times. The theory of D0-branes is known as matrix quantum mechanics.

In matrix quantum mechanics, there are nine matrices, each representing the dynamics of the D0-branes. The D0-branes have a potential energy made up of these interacting matrices that do not commute with each other, which encodes the uncertainty in the spatial dimensions. What I find most interesting about matrix theory is that many formulations of quantum gravity point to the same theory. This coincidence of a quantum theory with no a priori spatial dimensions is a fertile ground to investigate a pre–big bang phase where our smooth, classical, and expanding universe may have emerged from a reality in which space itself did not exist. At this impasse, I leave this speculation of a working emergent cosmological framework as a future project perhaps successfully pursued by a brilliant young cosmologist.

15

THE COSMIC MIND AND QUANTUM COSMOLOGY

Several years ago, a handful of scientists and I received an invitation to have a discourse with Dr. Deepak Chopra at the Carnegie Institution. The other physicist attending was Nobel laureate Frank Wilczek, so I couldn't resist the chance to be in great company. However, I approached the interview with the fear of being further sequestered from the community of scientists—many who held Chopra to mix mysticism with physics. Chopra and his colleagues have been proponents of the idea that consciousness creates the physical universe.[1] It is well known that certain scientists have debated with Chopra and even criticized his ideas as "out there."

Chopra is a brilliant medical doctor and effective communicator of science to the masses, and I was silently impressed and inspired by his willingness to debate unconventional ideas with renowned scientists. But he was clearly an outsider to the enterprise of research scientists. I was sympathetic to the resistance he received because he was not a member of the club, made claims that could not—yet—be experimentally supported, and was considered to be a charlatan by a handful of vocal, skeptical voices. And I always enjoyed watching

Chopra work the crowd and effectively debate with other scientists. Walking into my interview, I knew that he would certainly ask me about consciousness, quantum physics, and a topic that he has even published on recently with others, called cosmic consciousness.

In the middle of my interview, Chopra dropped the "C-bomb" on me. With his resonant baritone voice he asked, "So, Stephon, do you think the big bang came from consciousness?"

Like a skilled coward, I dodged his question. I said, "I'll take off my scientist hat and put on my Stephon-as-a-person hat and say that . . ." What I wanted to say was: "Deepak, I got into physics because I wanted to understand how this basic, direct experience of consciousness was connected to the fabric of the universe." Instead, I gave a weak response and said, "I think that a future generation of brilliant physicists should be brave enough to tackle the question." I chickened out and Chopra knew it.

I wasn't alone. At that time I was unaware that a handful of the architects of quantum mechanics, including Bohr, Heisenberg, Schrödinger, Wigner, and von Neumann, had been influenced by the question of consciousness in the physical world. Schrödinger was especially influenced by the work of German philosopher Arthur Schopenhauer and the Vedas, which posited the existence of a universal mind that contains all individual minds and the physical universe. Combining those with his own work, Schrödinger imagined the quantum wave function to be part of an undivided cosmic whole.

We have explored the consequences for our universe of the quantum, invariance, and emergence principles. The laws that follow from these principles, general relativity and quantum field theory, precisely predict a universe that expands from the big bang into a seemingly structureless early universe that vibrates with sound waves of energy and ultimately grows into the web of stars, planets, and galaxies. We don't have a complete understanding of the bang, the cause of the waves, and the emergence of space-time itself. And as we've seen, if we want to transcend the big bang singularity, we may very well need ideas that go beyond our current principles.

And those might not even be the biggest issues we face. The expansion of the early universe linked with the flow of entropy necessary for biological life is a hint at a deeper interdependence between life and the quantum universe. Did life emerge in the cosmos through a series of accidental historical events? Is there a deeper principle beyond natural selection at work that is encoded in the structure of physical law? And on top of that, the question that bothered Schrödinger and that got me into science in the first place: What is the relationship between consciousness and the fabric of the universe?

I am fully aware that I risk being written off as an oddball crank by the positivists in the room, because answering these questions might call into question the idea that the world out there is independent of us being there. But I must eat my own words to Deepak Chopra and embrace the stigma of weaving together theoretical ideas to entertain that question. In this final chapter, I am going to engage in an exercise of theoretical dreaming—into the principle of blackness. I am now going to pursue a speculation that intimately relates the big bang to the most complex entities that emerged from the universe: human beings endowed with conscious awareness.

This begins with the elephant in the room, the measurement problem in quantum mechanics. Quantum systems exist in a superposition of states until a measurement collapses the wave function into one unique state. In his book *Mathematical Foundations of Quantum Mechanics*, John von Neumann, the father of the modern computer, proved that when a quantum system exists in a superposition of states, a chain of measurements ultimately leading to the consciousness of an observer is what collapses the wave function into one definite state. He argued that this collapse by consciousness cannot be consistent with the mathematical framework of quantum mechanics, especially the linearity of quantum mechanics. This interaction between consciousness and quantum mechanics has to happen outside the constructs of quantum theory itself—unless consciousness is part of quantum physics from the start. We'll get back to this.

As we have discussed, in his book *What Is Life?* Schrödinger makes three key observations about what differentiates living from nonliving matter subject to the laws of physics including quantum mechanics. First, he predicted the basic helical structure for DNA using ideas from quantum mechanics that describe the periodic lattices found in metals. Second, he argued that living things fight against entropy, otherwise known as negentropy. Both predictions inspired the next generation of biochemists and biophysicists and continue to be foundational. And third, Schrödinger speculates about what it means that some (if not all) biological life has consciousness. He asks what is at first glance a strange question: Why are there many minds, each having their unique conscious experiences?

When Schrödinger was writing this work eighty years ago, he recognized that his era lacked a scientific account of consciousness, and so he resorted to philosophical and metaphysical ideas from Arthur Schopenhauer and Vedic philosophy. Both sources believe that the universe and all that occupies it carries varying forms of awareness. This view is often known as panpsychism. Panpsychism posits that consciousness is an intrinsic property of matter, the same way that mass, charge, and spin are intrinsic to an electron. So according to panpsychism, the electron and all substance come equipped with their own internal protoexperience of being an electron. This might sound crazy. Definitely there's a question about how an entity, say an electron, can have its own internal experience without resorting to an electron brain. The answer requires new physics or a fresh perspective on known physics. And we find a clue from African philosophy.

While there was an absence of African philosophy in my formal education, my musical colleague, legendary bassist Melvin Gibbs, introduced me to a view of the creation of the universe presented by the Bantu-Kongo people of West Africa that predates our modern big bang theory—but it has more, including an extra conceptual key to help us understand how to relate the quantum, cosmos, and consciousness with one another. In the Bantu cosmology the universe started in a state of nothingness called *mbungi*. Here nothingness includes the absence of space and time. Physical objects, such as particles and fields, usually exist in space-time. So *mbungi* is a prephysical state that is divided into what manifests as the physical, spatiotemporal world and a universal consciousness.[2] In the state of *mbungi* both the physical and conscious awareness are complementary, and have close semblance to the yin and yang in Taoism. Therefore, *mbungi* finds a natural home in quantum complementary in the context of the entire universe. Translating this into the language of cosmological physics, we can define a pregeometric universe as a quantum state that contains the potentiality of space-time and a fundamental form of consciousness as complementary pairs. To make a cosmological complementarity concrete we will need to actually have a formalism for quantum cosmology—a wave function of the universe.

In 1985 Stephen Hawking and James Hartle published a paper entitled "Wave Function of the Universe," which implemented the Schrödinger equation associated with quantum gravity known as the Wheeler-DeWitt equation. Unlike the original Schrödinger equation, which gives the time evolution of the wave function, the Wheeler-DeWitt equation is timeless.[3] Hartle and Hawking found a wave function of the universe, which is now famously called the Hartle-Hawking state.

The wave function of the universe is not some theoretical playground; it actually corresponds to the quantum state that underlies cosmic inflation that we discussed in an earlier chapter. In fact it was shown by cosmologist Alexander Vilenkin that the wave function of the universe can undergo a process of quantum tunneling from a state of nothingness—where space vanishes—to an inflating space-time.[4] Recall that a quantum system can go through barriers that are forbidden by classical physics. In this case the quantum universe can tunnel from a state of no-space, which is inaccessible to classical physics, into an inflating space-time. Because the Hartle-Hawking-Vilenkin state is connected to inflation, it is taken quite seriously by the cosmology community as a benchmark for doing calculations that correspond to satellite observations. Despite its pragmatic importance, the Hartle-Hawking-Vilenkin wave function presents a deep conceptual problem about the nature of time and the emergence of space at the big bang. Near the end of his life Hawking stated, "Asking what came before the Big Bang is meaningless . . . because there is no notion of time available to refer to. . . . It would be like asking what lies south of the South Pole."

Renowned Stanford University quantum cosmologist and co-inventor of inflation Andrei Linde gives us a hint of how to resolve the problem of understanding how the universe emerged from a state of no-space. Linde focuses on the fact that the Wheeler-DeWitt equation of quantum cosmology is timeless, which implies that the universe is "dead." Linde proposes the way out of this conundrum is to link consciousness with space-time. In a remarkable article Linde asks, "We cannot rule out the possibility a priori that carefully avoiding the concept of consciousness in quantum cosmology constitutes an artificial narrowing of one's outlook. . . . Is it not possible that consciousness, like space-time, has its own intrinsic degrees of freedom, and that neglecting these will lead to a description of the universe that is fundamentally incomplete?"[5]

But given this hint, how does the universe as we know it get realized at the big bang? This question is still up for debate, and I will argue that the wave function of the universe undergoes self-observation, a form of cosmic protoconsciousness, in the spirit of how the physical world emerges from *mbungi*.

The question, then, is how is it that this cosmic protoconsciousness can be timeless and spaceless? Explaining it seems difficult: neuroscience doesn't even have a complete explanation for where human consciousness comes from. My colleague David Chalmers is a leading researcher in the study of consciousness and pioneered a concept that he called the "hard problem of consciousness." In a nutshell, while neuronal activities of the brain can correlate with various perceptions and conscious experiences, it cannot explain our private, subjective, internal experiences of perception, self-awareness, emotions, and other experiential states of consciousness. The point the panpsychists make is that maybe neuroscience has set its sights too low. The Vedic system posits that the universe comes with its own singular consciousness, known as Brahman. With this concept, Schrödinger was able to answer his question about the multiplicity of individual minds if there was one universal mind by imagining that the multiplicity of individual minds are actually a hall of mirrors reflecting the one universal mind. For years, I was fascinated by this claim but didn't understand how many minds could be equivalent to one. And Schrödinger resisted making any further physical connections to his conviction about minds in the universe. What he failed to realize is that the fundamental principle of quantum mechanics, superposition, can hold the key to putting physical meat on the bone of his conviction. At the end of the day, fourteen billion years of cosmic evolution results in beings like you and me, endowed with the faculties of perception and consciousness. Is the question of consciousness solely a matter of emergence of the happenings of the brain? Or did the early universe encode the inevitability of conscious experience for specific cosmic functions?

The iconic theorist John Archibald Wheeler asked similar questions and believed that quantum mechanics plays a pivotal role in relating conscious observers to the very operation of the universe. While it may seem absurd, this type of reasoning led Wheeler to formulate a more bizarre version of the double slit experiment that he called the delayed choice experiment. Maybe the wildest thing about the delayed choice experiment is that it has been experimentally observed.[6] Here is an easy way to understand the experiment. Let's revisit the double slit experiment. We know that observing the electron before it enters the slit results in particle behavior, and not observing causes the electrons to behave like waves, due to interference. But what if we take the observation device away from the slits and place it at the screen? With precision technology, experimenters were able to wait till the last moment before the electrons hit the screen to watch it. In this case the electrons went through the slit like waves, and right before hitting the screen they collapsed like particles—the pattern was the same as when the double slit experiment measures the electron at the slit, which is not a wavelike pattern, even though you might expect that we'd get a wavelike distribution. From this bizarre behavior it means that the electron retrocausally went back in time and went through the slits like a particle. In other words, observers can delay their measurement of the electron's particle or wavelike behavior before it hits the screen.

In his own words, Wheeler asked, "So what does the quantum have to do with the universe? Perhaps everything, because in any fundamental theorem of existence, the large and the small cannot be separated." Here we get a sense of what it means for the universe to have a wave function. Like a quantum particle that traverses many paths from beginning to end, our quantum universe traces out many histories simultaneously from the bang to the present. In his idea that he called the participatory universe, which is a delayed choice experiment for the universe, Wheeler proposed that when conscious observers make quantum measurements of the early universe, we collapse the universe's quantum wave function to a history consistent with our existence.

As a younger, wide-eyed grad student I came across Wheeler's crazy ideas, such as "it from bit" (about the relationship between information and matter) and the absorber theory (about the direction of time) that he coauthored with Richard Feynman. Ironically, and similar to the experience Michael Faraday had with the idea of fields, these ideas began as outlandish but are now the norm of theoretical physics research, especially our quest to build a quantum computer. Nevertheless, most physicists shy away from entering Wheeler's rabbit hole of the participatory universe and exploring the possibility that seemingly insignificant specks like us can have any connection to and influence on the cosmos.

Wheeler believed that the universe implemented Darwinian evolution—and, as I and Salvador Almagro-Moreno argued, the entropocentric principle to create conscious life for the universe to observe itself. This self-measurement solves the measurement problem in quantum cosmology in one sense, because it gives a mechanism to collapse the universe's wave function to its current and future state of existence. However, the universe had to wait fourteen billion years for the first form of biological life to come on the scene. If there was no life before then, how could the universe measure itself? I do not know if Wheeler was aware of Schrödinger's musings on consciousness in the universe, but his statement points to the underlying physics: "So what does the quantum have to do with the universe? Perhaps everything, because in any fundamental theorem of existence, the large and the small cannot be separated."

This statement points to what I believe is the key insight to reconcile the lack of life as we know it to collapse the wave function of the universe—a nonlocal conscious observer. This nonlocality is complementary to locality in the same way the electron's position is complementary to its momentum. Recall that we discussed that the entirety of the quantum electron relies on opposing local and nonlocal properties; it is both a wave and a particle at the same time. Let us assume that consciousness, like charge and quantum spin, is fundamental and exists in all matter to varying degrees of complexity. Therefore consciousness is a universal quantum property that resides in all the basic fields of nature—a cosmic glue that connects all fields as a perceiving network.

Others are pursuing similar questions. Recently, Johnjoe McFadden and others have published neuroscience research arguing that consciousness is carried by electromagnetic field patterns distributed throughout the brain. Different organizational properties of fields can carry an array of conscious experiences. They point out each neuron in the brain can generate electric fields around their cell membrane and these individual fields can superpose across billions of neurons creating a complex pattern, rich in an organizational property, discovered by neuroscientist Giulio Tononi, called integrated information. Many prominent neuroscientists, such as Christof Koch and Tononi, believe some form of panpsychism is necessary to address the hard problem of consciousness.

What these proposals don't do is consider whether consciousness must be limited to our bodies. Our conscious experience is local in that our internal experience is in reference to our localized body in space and time. We might take this for granted, but that doesn't mean the universe does. What would it look like to have a nonlocal conscious experience? It could not be in reference to a place. According to the superposition principle in quantum mechanics, we could represent the nonlocal state of consciousness by superposing a large number of local conscious observers. If the universe's quantum state is endowed with a nonlocal state of consciousness, then according to this type of complementarity, it is dual to a superposition of local consciousness. This would resolve Schrödinger's paradox—why do so many minds have their own conscious experience? All these local minds need to be superposed to reconstruct a unitary, nonlocal, cosmic mind, like the positions of the electron needs to be superposed to create its field. The cosmic mind is contained in the local minds, though hidden from our everyday local experiences. What this means for you and me is that our consciousness contains an aspect of the cosmic mind. In Vedic philosophy this is referred to as the Atman, or the self.

This conclusion to some might be awe-inspiring or preposterous. When I set out to write this book, there was no way I could have imagined presenting this argument. If you find this line of reasoning preposterous, it is even crazier that we came into being to even be able to ponder these questions. Giving a shout-out to the distinguished Indian physicist: No one ever died from theorizing!

ACKNOWLEDGMENTS

When I first entertained the idea of writing this book I was overwhelmed with self-doubt. I want to especially thank K.C. Cole and Maria Popova for empowering and supporting me with some important tools and encouragement to get through writing this book. I also want to thank Mark Gould, one of the world's most brilliant social theorists, for his ongoing support and collaboration on theorizing the sociology of science with me. To my editor, the magician TJ Kelleher—thanks for yet another amazing journey. Thanks to Lara Heimert, Kelly Lenkevich, Sharon Kunz, Liz Wetzel, and the Basic, Perseus, Hachette team for helping to bring this book into reality. Thanks, Brandon Ogbunu, for your inspiration, guiding me through the very first stages of writing, and continuing to be a soundboard all the way through. Thank you Glenn Loury for writing *The Anatomy of Racial Inequality* and for your constructive criticisms that made this book stronger. Thanks to David Spergel for reading and providing scientific guidance on the manuscript. Thanks, Indradeep Ghosh, for your frank insights and guidance, Joao Magueijo, for many discussions, and Jaron Lanier, for co-writing a chapter and teaching me how to think outside the box. Finally, thank you to members of the Alexander Theory Group at Brown University.

Thank you Jerome Alexander, Salvador Almagro Moreno, Asohan Amarasingham, Sarah Bawabe, Willis Bilderback, Brown University, Will Calhoun, Liam Carpenter-Urquhart, Saint Clair Cemin, Colin Channer, Dwayne Ray Cormier, Everard Findlay, Batia Freedman-Shaw, Ashok Gangadean, Melving Gibbs, Heather Goodell, Jeff Greenwald, Leah Jenks, Ned Kahn, Brian Keating, Dagny Kimberly, Jaron Lanier, Janna Levin, Evan McDonough, Fernanod Pezzino, Vernon Reid, David Rothenberg, Susan Sharin, Jim Simons, Lee Smolin, Richard Snyder, Greg Tate, Greg Thomas, and Eric Weinstein.

NOTES

Chapter 2: The Changeless Change

1. Mathematically we can define a space-time vector as $x^{\mu} = (ct, x, y, z)$ where c is the speed of light and has dimensions $\frac{length}{time}$. This gives the time axis a dimension of length since the dimensions of the time interval, $ct = \frac{[length]}{time}[time] = [length]$, is length.

2. A local, freely falling observer in an exactly homogenous gravitational field experiences no forces. However, in a realistic situation there are gravitational field lines that are inhomogenous around the observer that will yield tidal forces.

3. Our direct experience of being in rotating environments seems different from being at rest, especially since rotation is a form of acceleration, which seems to mimic the effect of a force. We will soon see that this was part of Albert Einstein's insight about the equivalence of acceleration and being subject to a gravitational field without accelerating.

4. A geodesic can be understood more geometrically with vectors that are tangent to a point on a curved space, called a tangent vector. If we take a tangent vector, $\frac{dx^{\nu}}{d\tau}$, and transport it to a nearby region while keeping it parallel to itself, this amounts to satisfying the geodesic equation:

$$\frac{dx^{\nu}}{d\tau}\nabla_{\nu}\frac{dx^{\mu}}{d\tau} = 0$$

5. A solid example of a dynamical equation is that for how a pathogen like a virus grows in time:

Rate of growth of virus at a later time = (R-1)amount of virus at earlier time

Or symbolically

$$\frac{dy(t)}{dt} = (R-1)y(t)$$

Here R is famously known as the reproduction number. As I write this the COVID-19 virus has an average R of 2.3 in the United States. This leads to a solution of the number of infections to be exponentially growing in time $y = e^{ct}$.

6. All four Maxwell equations are contained in one equation if we write it in a manifestly Lorentz covariant form: $\partial_\mu F^{\mu\nu} = j^\nu$. The left-hand side of the equation is the four-dimensional derivative of the field strength tensor that contains all electric and magnetic field information, and the right-hand side is the four-dimensional current, containing the electric charge and currents.

Chapter 3: Superposition

1. The concept of phase space was pioneered by Ludwig Boltzmann, Henri Poincaré, and Josiah Willard Gibbs. Phase space is a key tool in formulating thermodynamics and statistical mechanics and in identifying attractors in chaos theory.

2. The particles, like the photon, that communicate the forces between matter are called bosons and have integer spin. But half-integer spin belongs in Alice's Wonderland. To get a feel, consider a child riding a merry-go-round. If the merry-go-round had half-integer spin, a child would have to go around two full rotations to ap-pear at his or her original position. Likewise, an electron needs to spin not 360 but 720 degrees to come back to its original spin orientation. M. C. Escher's drawing of ants traversing a Möbius strip is an example of a space that requires two orbits to return to the same place. Another fundamental characteristic of an electron is that it never stops spinning.

3. I even used the concept of quantum entanglement in my role of science adviser for Disney's *A Wrinkle in Time*, directed by Ava DuVernay.

4. Wave functions are vectors that live in a complex vector space called a Hilbert space. Said another way, a Hilbert space is a space of all possible wave functions and has all the usual properties of linear vector spaces commonplace to linear algebra.

Chapter 4: The Zen of Quantum Fields

1. *Kensho* is a Japanese word that translates to what Westerners call enlightenment, a liberating experience recorded by Siddhartha Gautama, the historical Buddha. According to the Buddha this potential to experience *kensho* is accessible to all humans. Similar accounts of the enlightenment experience were recorded by individuals across different cultures. Meister Eckhart calls it a breakthrough. The theologian St. Symeon calls it waking up. "I wasn't there." "There was only the tree." "A sense of utter liberation and bliss." "It is overwhelmingly positive." "It's like being drunk, but on reality." "It's more real than real."

2. This cosmic ocean is called the vacuum state of quantum field theory. The vacuum state can both create and annihilate particles with quantum operators called creation and annihilation operators, respectively.

3. This "simple way" is a linear interaction between the photon vector potential and the fermion current. Such a linear interaction guarantees a local phase invariance for the electrons provided the gauge field simultaneously compensates the phase with its own phase transformation.

Chapter 5: Emergence

1. In modern times, solid-state physics is often called condensed-matter physics, which also involves the study of the various states of matter, such as liquid, gaseous, and crystalline, where many quantum particles interact quantum mechanically.

2. Kenneth Chang, "When Superconductivity Became Clear (to Some)," *New York Times*, January 8, 2008, https://www.nytimes.com/2008/01/08/science/08super.html.

Chapter 6: If Basquiat Were a Physicist

1. The electron can only have discrete (quantum) spin about its axis of rotation and never stops spinning. This is to be contrasted with a macroscopic spinning object, which can continuously change its spin orientation from up to down, such as when a top starts spinning and eventually falls to its surface. If the electron's spin flips from up to down, it does so discretely, in a quantum unit of spin. And strangely, experiments seem to require one electron to coexist in a spin up and spin down state at the same time.

2. A conformal invariant theory is analogous to looking at features of a theory under a "magnifying glass." This is equivalent to zooming in or rescaling the coordinates of the observables of the theory. If the predictions of the theory do not change under this zooming in, or zooming out, the theory is said to be scale-invariant. The microcosm has the same properties as the macrocosm. In some special cases, conformal invariance is the same as scale-invariance.

3. Thomas Kuhn, *The Structure of Scientific Revolutions* (Chicago: University of Chicago Press, 2012).

4. Du Bois wrote that "the Negro is a sort of seventh son, born with a veil, and gifted with second-sight in this American world,—a world which yields him no true self-consciousness, but only lets him see himself through the revelation of the other world. It is a peculiar sensation, this double-consciousness, this sense of always looking at one's self through the eyes of others, of measuring one's soul by the tape of a world that looks on in amused contempt and pity. One ever feels his twoness,—an American, a Negro; two souls, two thoughts, two unreconciled strivings; two warring ideals in one dark body, whose dogged strength alone keeps it from being torn asunder." W. E. B. du Bois, *The Souls of Black Folk* (Orinda, CA: SeaWolf Press, 2020).

5. The conscience collective constitutes the shared social values within any social order; it regulates norms and activities in every social order, but it is not always well integrated; the values are not always consistent, and thus the social order might be poorly ordered. Here we might think about the conscience collective for both the larger society and for the community of physicists.

6. Phenomenologists sometimes call these cultural norms, which differentiate between sense and nonsense, the "lifeworld." If the lifeworld is

functioning effectively, it is tacit—we are unaware that we are seeing the world through it.

7. Durkheim recognized three functions of punishment. The first is the one usually discussed in the literature. Punishment creates an incentive to conform to social and cultural norms; it realigns what is in the interest of actors. Second, as noted in the text, punishment delineates the boundary of what is allowed. If a public violation of a normative expectation goes unpunished, it will undermine the clarity of the normative orientation; in contrast, the punishment of specific actions reinforces normative boundaries. Third is what we call the "sucker effect." If conformity to a normative expectation continually disadvantages an actor, this will undermine her commitment to the relevant norm; she will feel like a sucker conforming while those who violate the norm are advantaged over her. If she knows, instead, that violators are likely to be punished for their violations, this enables her to sustain her sense of obligation.

8. There is a huge amount of sociological literature on deviance and marginality. Robert Ezra Park coined the term "marginal man," but his usage was different than ours. Also relevant is Georg Simmel's discussion of "the stranger." Everett Hagen was one of the first to ask if deviance might be related to innovation. See his *On the Theory of Social Change: How Economic Growth Begins* (Homewood, IL: Dorsey Press, 1962), 573.

9. Eric Felisbret, "Legal Venues Celebrate Graffiti as an Art Form," *New York Times*, July 18, 2014, https://www.nytimes.com/roomfordebate/2014/07/11/when-does-graffiti-become-art/legal-venues-celebrate-graffiti-as-an-art-form.

Chapter 7: What Banged?

1. The expanding universe model was independently discovered by
    1. Alexander Friedmann et al., "Über die Krümmung des Raumes," *Zeitschrift für Physik A* 10, no. 1 (1922): 377–86.
    2. Georges Lemaître, "Expansion of the Universe, A Homogeneous Universe of Constant Mass and Increasing Radius Accounting for the Radial Velocity of Extra-Galactic Nebulæ," *Monthly Notices of the Royal Astronomical Society* 91, no. 5 (March 1931): 483–90.

3. H. P. Robertson, "Kinematics and World Structure," *Astrophysical Journal* 82 (November 1935): 284–301.
4. A. G. Walker, "On Milne's Theory of World-Structure," *Proceedings of the London Mathematical Society, Series 2* 42, no. 1 (1937): 90–127.

Chapter 8: A Dark Conductor of Quantum Galaxies

1. All observed galaxies have dark matter. See Jim Shelton, "New Studies Confirm Existence of Galaxies with Almost No Dark Matter," YaleNews .com, March 29, 2019, https://news.yale.edu/2019/03/29/new-studies -confirm-existence-galaxies-almost-no-dark-matter.

2. The spatial anisotropy in the density of baryonic and dark matter deviates by approximately one part in ten thousand from the average homogeneous and isotropic cosmic microwave background thermal bath.

3. It was shown by Kris Pardo and David Spergel that it is difficult to reproduce the effects of dark matter in the CMB fluctuations in the early universe with MOND. See Kris Pardo and David Spergel, "What Is the Price of Abandoning Dark Matter? Cosmological Constraints on Alternative Gravity Theories," *Physical Review Letters* 125 (October 19, 2020): https://arxiv.org/abs/2007.00555.

4. To be precise, dark matter is five times the density of baryonic (visible) matter.

5. Kaplan is known for producing the critically acclaimed film about the Large Hadron Collider, *Particle Fever*, which tells the story of the vision of the LHC and the physicists behind it. Coincidentally, I was the moderator for the world premier for the film at the NeueHouse in New York City.

6. Classical field theories have conserved quantities. For example, in electromagnetism the current is conserved. A quantized theory should also conserve all classical currents. However, there are some currents that are broken upon quantization. The amount of current that is broken is called an anomaly. However, a healthy quantum theory fixes these anomalies by realizing that topology needs to be considered in the quantum theory so as to reinstate the current conservation.

Chapter 10: Embracing Instabilities

1. The uncertainty principle can be conveniently written as $\Delta x \sim \frac{1}{m\Delta v}$, the uncertainty in the position of the particle in a harmonic oscillator is bounded by the size of the system. Therefore, the velocity can never be zero according to the uncertainty relation above.

2. General relativity provides an equation that relates energy and matter to the curvature of space-time. The configuration of the matter and energy should warp the space-time field in a specific dynamical manner dictated by the Einstein field equations similar to how a magnet emanates and bends a magnetic field surrounding it. Microscopically, a magnet emerges from interacting atomic spins in a metal. The cosmological constant interacts with the space-time in a manner that causes it to accelerate in its expansion without bound.

3. The Ashtekar variable is similar to QCD in that the dynamics of both theories are encoded in a gauge field connection. In QCD the connection enjoys a larger symmetry [SU(3)] than that Ashtekar connection, [SU(2)].

4. Stephon Alexander and Raúl-Rubio, "Topological Features of the Quantum Vacuum," *Physical Review D* 101, no. 2 (2020); Stephon Alexander, Gabriel Herczeg, and Jinglong Liu, "Chiral Symmetry and the Cosmological Constant," *Physical Review D* 102, no. 8 (2020).

Chapter 11: A Cosmologist's View of a Quantum Elephant

1. I first spoke to Sir Roger Penrose, the inventor of twistor theory, who encouraged me to discuss the idea with Ashtekar. Twistors are maps from events in space-time to the celestial sphere that has close semblance to the symmetries enjoyed by the Ashtekar connection. I first thought that my idea behind parity violation and quantum gravity had to do with twistors.

Chapter 12: The Cosmic Biosphere

1. "John von Neumann Compares the Functions of Genes to Self-Reproducing Automata," HistoryofInformation.com, http://www.historyofinformation.com/detail.php?id=682.

2. B. Jesse Shapiro et al., "Origins of Pandemic *Vibrio cholerae* from Environmental Gene Pools," *Nature Microbiology* 2, article no. 16240 (2017).

As Salvador Almagro-Moreno communicated to me, "Virulence adaptive polymorphisms (VAPs) circulate in environmental populations and must be present in the genomic background of a bacterium before it can emerge as a successful pandemic clone."

3. The entropy of a black hole is calculated to be $S = \frac{4\pi k G M_{BH}^2}{\hbar c}$.

4. Ludwig Boltzmann, "The Second Law of Thermodynamics," in *Theoretical Physics and Philosophical Problems*, ed. B. F. McGuiness (New York: D. Reidel, 1974).

5. One potential loophole to the low-entropy initial state is that there was a previous state what was actually even more entropic that reduced its entropy. For example, if the big bang started out dominated with black holes, then they could evaporate and produce the CMB radiation. If the massless radiation, such as photons, got homogeneously distributed then decayed into matter, then we would have a lower-entropy situation. However, the original entropy presumably in the form of black holes would have to get diluted. This idea can perhaps be implemented in a mechanism developed by Peter Mészáros and me where we postulated that primordial black holes that form after inflation could explain the CMB and dark matter if they undergo Hawking evaporation.

6. Fred C. Adams et al., "Constraints on Vacuum Energy from Structure Formation and Nucleosynthesis," *JCAP* 03 (2017).

Chapter 13: Dark Ideas on Alien Life

1. A similar phenomenon happens with electrons and magnetic fields in the fractional quantum Hall effect.

Chapter 14: Into the Cosmic Matrix

1. Arvind Borde and Alexander Vilenkin demonstrated that even cosmic inflation is geodesically incomplete and does not escape the cosmic singularity. See Arvind Borde, Alan H. Guth, and Alexander Vilenkin, "Inflationary Spacetimes Are Incomplete in Past Directions," *Physical Review Letters* 90, no. 15 (April 15, 2003).

2. To be precise, if x and p are position and momenta variables then quantization rules promote their Poisson brackets (curly brackets) to commutation (square brackets) $\{x, p\} \rightarrow i\hbar[x, p]$. Classical dynamics of an observable is given by Hamilton's equation, which states that the time evolution is generated by the Poisson brackets of the Hamiltonian and the observable of interest. For example, the classical time evolution of the position of a particle, x(t) is given by the Poisson bracket between the Hamiltonian and the position.

$$\frac{dx}{dt} = \{x, H\}$$

Whereas the time evolution of the position operator is given by the commutator between the Hamiltonian and position operators.

$$\frac{d\hat{x}}{dt} = -i[\hat{x}, \widehat{H}]$$

3. Here we are assuming that the spatial dimensions have the topology of circles whose local products are tori. These topologies admit one-cycles that strings can wind around without collapsing to a point.

4. This initial condition for string cosmology is an extension of the Copernican principle in the standard big bang, which states that there is no preferred vantage point in the universe. The Copernican principle is consistent with assuming homogeneity and isotropy not only in space but in the degrees of freedom that occupy the universe.

5. A cyclic BV mechanism was studied by Brian Greene, Daniel Kabat, and Stefanos Marnerides. They found that the BV mechanism can exhibit a cyclic cosmology if the theory incorporates higher derivative terms in the gravitational theory.

6. In a closed string theory we can add open strings that satisfy both von Neumann and Dirichlet boundary conditions.

$$n^a \partial_a X^\mu = 0, \ \ \mu = 0, \dots p$$

The above equation states that the derivate of the string ends on p space-time dimensions. This is a Dirichlet boundary condition. This p-dimensional hypersurface defines the worldvolume of a Dp-brane.

$$X^\mu = 0, \qquad \mu = p + 1, \dots 9$$

7. The above equation states that the string vanishes in the other dimensions.

Chapter 15: The Cosmic Mind and Quantum Cosmology

1. Deepak Chopra, MD, and Menas C. Kafatos, PhD, *You Are the Universe: Discovering Your Cosmic Self and Why It Matters* (New York: Harmony Books, 2017).

2. Kimbwandende Kia Bunseki Fu-Kiau, *African Cosmology of the Bantu-Kongo: Principles of Life and Living* (Brooklyn, NY: Athelia Henrieta Press, 2001), 17–54.

3. In general relativity the Hamiltonian is constrained to be zero. As a result the associated Schrödinger equation will no longer have time dependence. The state of the universe will therefore be timeless, which means that the universe as a whole is changeless. In order to obtain time dependence, one has to introduce an external clock relative to the rest of the universe, and matters of this sort are called the problem of time in quantum gravity.

4. Alexander Vilenkin, "Creation of Universes from Nothing," *Physics Letters B* 117, no. 1–2 (November 1982): 25–28.

5. Andrei Linde, "Universe, Life, Consciousness," https://static1.squarespace.com/static/54d103efe4b0f90e6ca101cd/t/54f9cb08e4b0a50e0977f4d8/1425656584247/universe-life-consciousness.pdf.

6. Vincent Jacques et al., "Experimental Realization of Wheeler's Delayed-Choice Gedanken Experiment," *Science* 315, no. 5814 (February 16, 2007): 966–968, https://arxiv.org/abs/quant-ph/0610241.

INDEX

Stephon Alexander is a professor of physics at Brown University and the 2020 president of the National Society of Black Physicists. He is also a jazz musician and released his first electronic jazz album *Here Comes Now* with Erin Rioux and *God Particle* with bassist Melvin Gibbs. The author of *Jazz of Physics*, Alexander lives in Providence, Rhode Island.